RAILWAY SIGNAL ENGINEERING
IN THE
MECHANICAL ERA

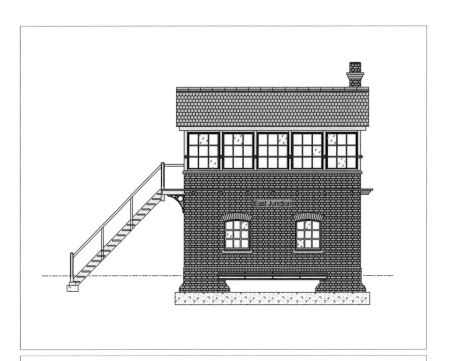

RAILWAY SIGNAL ENGINEERING
IN THE
MECHANICAL ERA

ORIGINALLY TITLED
RAILWAY SIGNAL ENGINEERING (MECHANICAL)
by

L. P. Lewis

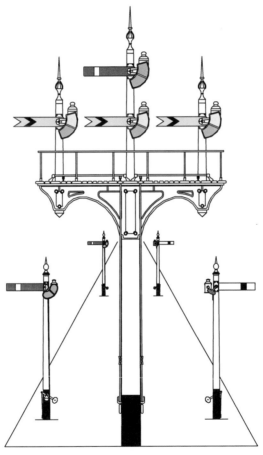

REVISED AND RE-ILLUSTRATED IN COLOUR
by
Gordon Roberts

RENASCENT BOOKS

□ Heritage Railway Signalling Series □

PUBLISHED BY

TGR Renascent Books
27 Springdale Court
Mickleover, Derby DE3 9SW
United Kingdom
2013

ISBN 978-0-9563585-8-5

www.renascentbooks.co.uk

Printed and bound by
CreateSpace,
Charleston, South Carolina,
United States of America

For Elizabeth,

whose constant support and unfailing enthusiasm
made this book possible.

CONTENTS

CONTENTS

LIST OF ILLUSTRATIONS

Chapter 9—Ground Frames

Chapter 10—Signalling Schemes

PREFACE

THIS standard treatise on mechanical railway signalling by Leonard Lewis was written at the turn of the twentieth century. Originally published in 1910 as *Railway Signal Engineering (Mechanical)*, a second edition followed in 1912. A third edition, revised and enlarged by J. H. Fraser, appeared in 1932. Since its original publication, now more than 100 years ago, much if not all of the mechanisms and practices described and illustrated have disappeared from the modern high-speed railways of Britain and the rest of the world. In his preface to the first edition, Lewis wrote that he intended the book to be '... suitable for men who are engaged in railway work, but not necessarily in connection with the Signalling Engineer's Department.' Today, such men no longer have any professional interest in what to them is now archaic and superseded.

However, with the popular growth of preserved heritage railways, and the dedicated reconstruction and re-creation of many railway artefacts by enthusiasts, it is no longer possible to state categorically that any particular mechanism or operating procedure described in these pages is extinct. Although they may have disappeared from modern railways in the electronic and computer controlled age, original or replica items or otherwise obsolete methods of working may well be in regular use on preserved branch line railways or be on display in railway museums.

Herein lies the main inspiration for this new edition at the start of the twenty first century. Lewis's book, once describing the very cutting edge of railway technology, has become with the passage of time a valuable work of history. Nevertheless, its contents may still be very relevant and of inestimable value to those responsible for the maintenance and operation of precious and irreplaceable signalling equipment on preserved steam and diesel railways, wheresoever those lines might be. Again, the ever growing band of collectors and restorers of old signalling equipment will find the technical material in these pages of more than passing interest. Likewise, enthusiasts viewing the artefacts on display in railway museums might find this volume can usefully supplement the information provided in simplified guide books and explanatory leaflets.

Railway Signal Engineering (Mechanical) is long out of print. The present derivative work is based on the 1932 edition and non of Lewis's original text, nor that later added by Fraser, has been omitted from this reprint. It is in every word as the original, except for a few minor corrections and one important detail. That is, the captions to some of the drawings have been amended to more accurately reflect the intent of the illustration, than did Lewis's original captions. Also note that no illustrations have been omitted, although a few have been added. However, as the most cursory glance through these pages will show, all the illustrations have been redrawn, in many cases substituting more realistic depictions of signals and mechanisms for the sometimes rather crude sketches in the original. Most notably, colour has been used, not only to provide a more visually appealing book for the enthusiast and the historian, but also in the hope that it adds somewhat to the understanding of technical descriptions and of the illustrations themselves.

G. ROBERTS (2013)

PREFACE TO THE FIRST EDITION

THE object of this book is to provide a brief outline of the general principles of Mechanical Railway Signal Engineering (excluding power-worked signals, etc.), with examples of their practical application, in a form suitable for men who are engaged in railway work, but not necessarily in connection with the Signalling Engineer's Department.

In order to assist readers who may not be familiar with the terms and names of apparatus employed in railway signalling most of the technical terms have been explained and the apparatus illustrated, as they occur, but in a few cases the description of apparatus is given later in the book; reference to the index will enable readers to find the description required.

No attempt has been made to describe every piece of apparatus employed on British Railways, nor yet to give the signalling practice of every railway company; as far as possible, however, the *average* practice of British railways is given.

Railway signalling practice in Great Britain has not yet been rigidly standardised, and there has been no serious attempt on the part of the various railway companies to adopt the same standard materials and apparatus.

Every railway company carrying passengers is subject to the regulations of the Board of Trade, and all important railways have adopted the Block Telegraph Rules of the Railway Clearing House (with slight modifications), so to that extent the main principles of British railway signalling have been standardised. In minor details each company employs its own particular practice.

In arranging the order of the chapters it has been thought preferable to describe the apparatus employed in signalling installations before giving examples of signalling schemes, so that when these are considered, the reader may better understand the working of the various articles which have to be indicated on the signalling plans.

L. P. LEWIS(1910)

PREFACE TO THE SECOND EDITION

BRITISH Railway Signalling Practice has not been subject to any fundamental change during the last decade. Great advance has been made in the installation of Track Circuits, and improvements made in their component parts.

The work of Standardising Signalling Material is being undertaken by the new Ministry of Transport, and the results of the Standardisation Committees may be expected to be published in due course by the Ministry.

It has not been necessary greatly to alter the text of the book for this edition, but the latest available information has been inserted.

L. P. LEWIS (1912)

PREFACE TO THE THIRD EDITION

THE fundamental principles of British Railway Signalling outlined in the first and second editions of this book have been little, if at all, altered by lapse of years. The detailed applications of those principles have, on the other hand, suffered a continual change under the influence of two great forces: the insistent call for economy; and the radical development of electrical devices and appliances. The former of these influences has been felt in a general movement towards consolidation and standardisation, resulting, firstly in the amalgamation of the railways of Great Britain into four large groups accompanied by a natural tendency toward uniformity of practice within each group; and secondly in the appointment of various committees to inquire into the possibility of standardising all classes of engineering materials and appliances. That these inquiries have had valuable and concrete results may be judged from the frequency of references in the present volume to "British Standard" designs.

Though many of the designs and practices which were formerly employed have, by these changes, been rendered obsolete for new work, the descriptions of them which occurred in the previous editions have, in general, been retained, as it is considered that this will make possible a truer appreciation of the principles which underlie present-day methods.

The important and ingenious contributions which electrical science has made to the practice of railway signal engineering within the last decade are rather outside the purview of this work, except so far as these developments have had appreciable repercussion on mechanical ideas and practice. But the essential unity of the mechanical and electrical branches is fully realised, and a volume is now in preparation which is intended to be the electrical complement of the present book.

The writer wishes to tender his acknowledgements to the Controller of H.M. Stationery Office for his permission to publish portions of the Ministry of Transport Requirements and to Messrs. The Westinghouse Brake and Saxby Signal Company for the use of illustrations of their Double Wire Apparatus.

J. H. FRASER (1932)

1
GENERAL IDEAS

RAILWAYS are provided with signals for the purpose of controlling the running of the traffic, and before considering what form these signals take, it is necessary to understand the system adopted for running trains.

When considering a length of railway, the first requirement is to ascertain in which direction the trains run.

If there is only one pair of rails, as a rule trains will run over that piece of line in either direction, but precautions have to be taken that two trains shall not start from opposite ends of the line to collide with each other. Where there is more than one pair of rails it is usual to have one fixed direction in which the trains shall run over a particular pair. In railway terms, a line (that is a pair of rails) is for traffic going in the UP or else in the DOWN direction.† This, of course, only refers to running lines, not to sidings. In cases of emergency it is sometimes necessary for a train to run on the UP LINE in the DOWN direction; this is termed WRONG LINE RUNNING, but before this can be done important rules have to be rigidly carried out to ensure safety.

At stations, movements often have to be made in the WRONG direction over a short portion of line, but, as a general rule, a train moving say in the UP direction is turned on to a line used for traffic in that direction as soon as possible.

Where there are several lines running parallel to each other the UP LINES may be placed together and the DOWN LINES together, or they may be arranged alternately UP LINES and DOWN LINES.

The general notation in Great Britain is that trains run on the left-hand side of the railway, and that a particular signal always refers to one line only, and never to more than one line. Before it is possible properly to understand present signalling practice it is necessary to glance at some of the early arrangements employed.

Whilst locomotives were employed in mineral working, and slow speeds were general, verbal instructions were ample to secure safety. With the increase of speed, and passenger traffic developing, it soon became evident that something more definite would have to be employed. As a first attempt, hand-signalmen were placed at important junctions to work the points and control the train running. The next advance was the use of a board, painted RED, fixed to a post. When a train was required to stop, the broad side of the board faced the driver, but when there was no necessity for stopping, the board was turned so that the edge of it faced the driver (Fig. 1).

This proved to be objectionable, as the indication for a clear line was the *absence* of the red board; hence should the board have been destroyed by accident a *clear* signal was given.

This was followed by the adoption of a distinctive signal for CLEAR as well as for DANGER.

The most common arrangement was the Disc and Crossbar signal (see Fig. 2). When the cross-

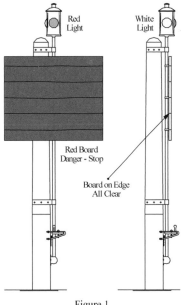

Figure 1
Red board signal

---

† It is the convention to refer to the line carrying trains in a direction *towards* London as the Up Line and that taking traffic *away* from London as the Down Line. Cross country lines often have the Up and Down direction decided with reference to an important town, station or junction situated to the east or west.

---

bar faced the driver it indicated DAN-GER, and when the crossbar was turned edge-on the disc faced the driver indicating CLEAR. A rotating disc signal used by the London and South Western Railway is shown in Fig. 3.

For night working a lamp was placed on the post with lights to show RED for DANGER and WHITE for CLEAR.

Besides telling a driver to stop or proceed, it was found necessary at a junction to inform him in which direction he was being sent.

At first the points were worked by hand levers fixed near them (Fig. 4), and when the lever lay to the right hand, the points were set to run a train in that direction; similarly, when the lever lay to the left, the points were set for the left. To aid the driver in picking out the lever, it was painted red. Another arrangement was the Point Capstan (Fig. 5).

When the points were worked from

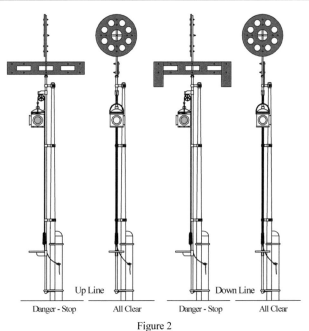

Figure 2
Disc and crossbar signals

levers some distance away, the idea of indicating the position of the points was retained by adopting an additional indicator lever working along with the points, termed a "Point Indicator." It was later fixed on a post, and there were coloured glasses and a lamp fixed to show White when the points lay for the main line, Red when they were not properly set in either position, and Green when set for the branch line (Fig. 6).

The arrangement for forwarding one train after another on the same set of rails was to allow a certain interval of time to elapse before the second one was forwarded.

A very common arrangement at a station was to have two discs one above the other: as soon as a train passed, the top one, painted RED, was turned to face any following train; at the end of five

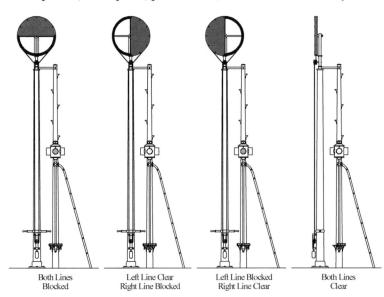

Figure 3
Rotating disc signals

minutes this board was turned on edge and the lower Green board was turned facing a following train, to indicate that such a train must proceed at caution; at the end of another five minutes both were turned to the clear position.

When the electric telegraph was invented it became possible to ascertain that a train had actually reached the station ahead before a second train was allowed to follow on the same set of rails. This arrangement was known as the SPACE

INTERVAL SYSTEM, the old arrangement being termed the TIME INTERVAL SYSTEM.

One objection to the adoption of the space interval system was the great distance between some of the stations, as very often it took a train *more* than five minutes to reach the station ahead, and as until it did arrive no following train could be sent, it was obvious that the carrying capacity of the line would be decreased unless the distances between stations were shortened.

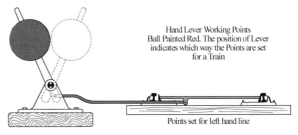

Figure 4
Point lever with painted handle

The line then had to be cut up into comparatively short sections (termed BLOCK SECTIONS) by means of small signal stations, which were generally termed BLOCK BOXES or BLOCK POSTS.

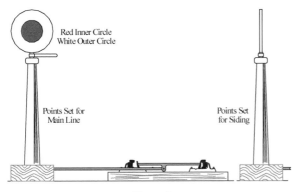

Figure 5
Point capstan

The usual equipment of a simple block box was a telegraph instrument for sending the requisite messages from one block box to the next, and outdoor signals for each set of rails, one signal being fixed near the box, termed the HOME signal, which trains were not allowed to pass when at DANGER, and a signal fixed further out (nearer an approaching train) termed a DISTANT signal, which the driver could pass at danger, provided he so got his train under control as to be able to stop short of any obstruction either before or at the Home signal. Home signals were usually semaphore signals (see Fig. 7), but it was quite common, even after the introduction of the semaphore signal, for the distant signal to be a disc signal.

The signal lights for junctions usually were Red for stop, Green for the branch line when clear, and White for the main line when clear.

Even under these arrangements accidents were not uncommon, and on lines where the space interval was not in operation accidents were still more numerous. At junctions accidents sometimes occurred owing to two conflicting signals being in the clear position at the same time, thus allowing two trains to collide.

This turned the minds of inventors to the problem of making it mechanically impossible for the signalman to have conflicting signals pulled to the clear position at the same time.

This was effected in two ways:

Interlocking the signal wires with the outdoor point connections.

Interlocking the signal levers with each other, and by interlocking the signal levers with the point levers.

These inventions made it impossible for a signal to be pulled to the clear position unless the points were first in the correct position for the train to proceed.

In 1889 an Act of Parliament was passed compelling all railway companies to adopt safety appliances and SPACE INTERVAL WORKING for passenger trains, and the Board of Trade

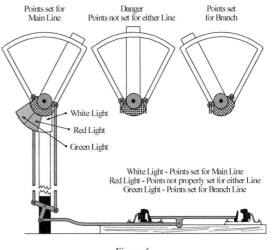

Figure 6
Point indicator

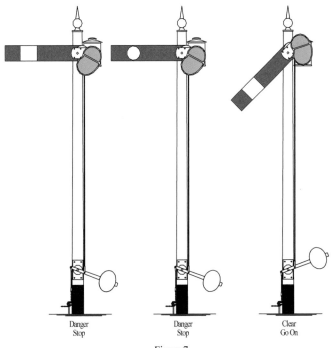

| Danger Stop | Danger Stop | Clear Go On |

Figure 7
Semaphore lower quadrant home signals

issued regulations stating certain requirements which would have to be complied with before new lines would be authorised for passenger traffic.

In 1919 an Act of Parliament was passed instituting a Ministry of Transport; and to it were transferred the functions of the Board of Trade affecting railway equipment and working. The Regulations mentioned above were, in 1925, re-issued in modified form as "Requirements for Passenger Lines, and Recommendations for Goods Lines, of the Minister of Transport in regard to Railway Construction and Operation," and, as these "Requirements" form the basis of modern British signalling, reference will be made to them from time to time throughout this volume.

The "Requirements" (copies of which can be obtained through any stationer, or direct from H.M. Stationery Office, price 1s.) are divided into four sections, as follows:

A. Documents to be furnished.
B. Requirements and Recommendations.
C. Modes of Working Single Lines.
D. Appendices.

A. Documents to be Sent to the Secretary, Ministry of Transport.

Paragraphs I to IX refer to Tables of Gradients, Earthworks, Bridges, Level Crossings, etc., etc., which must be submitted. Paragraph X refers to detailed information under clauses headed Permanent Way, Fences, Drainage, etc., etc., and includes under its eighth clause "Plans or Dimensioned Diagrams of the Signalling Arrangements at all junctions, stations, block posts, etc."

A note is made that *"The detailed information asked for under Paragraph X, Clauses 5, 7 and 8, should, whenever possible, without causing undue delay to the works, be forwarded in time to permit of examination by the Minister of Transport before works under this head are commenced. Before a statutory second notice is forwarded, the tabular statements under Paragraphs I to IX inclusive and the information under Paragraph X, Clauses 1 to 4 and 6 will be required".*

B. Requirements and Recommendations.

NOTE.—*These apply to construction or reconstruction and alterations or additions. It will be seen that references are made in the text to the possibility of relaxation to meet individual cases. It should also be noted that, in order to secure economy, with due regard to safety, when no references to relaxation are made, these Requirements may be modified at the discretion of the Minister of Transport, having regard to such special circumstances as may be submitted for consideration in each case. Standardisation of signalling and block working, &c, principles is also desirable. With these ends in view, it will therefore be desirable to submit, whenever practicable, plans of works, for which approval is required, before they are commenced.*

2
CLASSES AND USES OF SIGNALS

THE signals now employed for controlling the movements of trains can be roughly divided into two main classes, *viz*:

1. Signals which are not fixed in any one place but are exhibited only when required to control some particular movement of a train or engine. Such signals take the form of flags and lamp lights of different colours, Red, White or Green. They are chiefly used for shunting purposes, and are carried by the shunter, or shown out of the signal box window by the signalman when required. These shunting signals are sometimes assisted by whistles or blasts from a shunting horn.

2. Signals which are erected at some particular place are commonly called "Fixed" signals. These are so fitted as to be capable of exhibiting different indications, as required by the signalman, and usually take the form of SEMAPHORE signals or COLOUR LIGHT signals for the main lines, and small SEMAPHORES or DISCS for sidings.

THE SEMAPHORE

THE semaphore signal, with few exceptions has been the universal signal for main line working. It consists of a post on which an arm is fitted, capable of being shown with the arm horizontal or inclined.

In the case of the new British Standard Semaphore the arm moves upward (Fig. 8); while in older forms it moves downward, the universal ruling being, the arm horizontal indicating DANGER—STOP, and the arm inclined CLEAR—PROCEED.

As the Danger indication of both Upper and Lower Quadrant arms is the same, *i.e.*, horizontal, an intermixture of the two types is not considered undesirable, and is frequently met with in practice. For the sake of appearance, however, it is usual where more than one arm is fixed to the same post, for the arms to be of the same type.

In Great Britain the arm is generally about 4 ft. long and about 10 in. broad. Usually it is fitted with a spindle at one end, on which it swings. On a few railways the spindle was fitted at about the centre of the arm (the so-called somersault signal, see Fig. 9), but the indications are the same as when the arm is fitted with the spindle at the end. The arm is fitted to project on the left-hand side of the post.

The face of the arm (that is, facing drivers) is, in the case of Stop signals, painted RED with a band of WHITE. The reverse side of the arm is painted WHITE with a BLACK band across it.

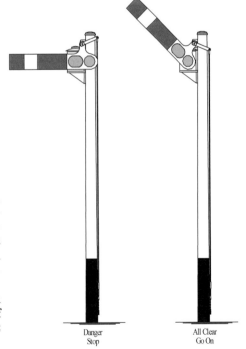

Danger
Stop

All Clear
Go On

Figure 8
British Standard upper quadrant semaphore

DISTANT SIGNALS

TAKING the main line signals as they are seen by a driver approaching an ordinary block box (see Fig. 10), the first signal is the Distant signal.

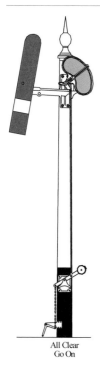

All Clear
Go On

Figure 9
*Balanced or
"somersault" signal*

This signal is a CAUTION signal, warning the driver, when it is in the horizontal position, that he may be required to stop at the next signal applying to the line on which he is running.

To distinguish this signal from a signal which must be treated as a STOP signal, and not passed at danger without the special authority of the signalman, the arm is coloured YELLOW with a BLACK band, and the end is shaped with a V notch (Fig. 11), while, at night, while the arm is horizontal a YELLOW light is shown. All STOP signals have square ended arms. The new Standard Distant signal has the bands on the arm in the form of a V to correspond with the notch in the arm. This is a great assistance in enabling the signal to be recognised at a distance when there is a bad background.

With high speed running it is im-

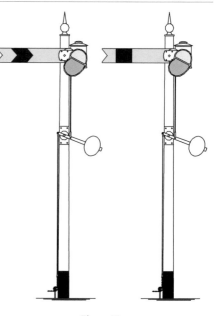

Figure 11
Distant signals

perative that a driver shall be informed in ample time if he is required to stop; accordingly the distant signal has to be placed some distance from the first STOP signal. The precise distance depends on many circumstances. The essential requirement is that the driver shall be warned in *sufficient* time to enable him to bring his train to a stand at the STOP signal without difficulty. The distance in which a driver can bring his train to a stand depends on the speed he is running at, the gradient of the line on which he is running, the state of the rails—whether greasy or dry—and the power of his brakes relatively to the weight of the train.

Trains fitted with either the automatic air or automatic vacuum brake, with brake blocks on every wheel, running at about 55 miles per hour have been brought to a stand less than 400 yards from the place where the brakes were first applied.

It is usual to place the distant signal about 1,000 yards out from the first STOP signal, where the line is approximately level, and ordinary express train running is the rule. With a falling gradient the distance should be greater, and may be 1,200 yards or more; with a stiff rising gradient the distance may be reduced.

Besides placing the distant signal at the correct distance from the first STOP signal, it is essential that the drivers shall have a good view of it for some distance before reaching it. The great importance of this will be recognised when it is remembered that half a mile of distance *may* correspond to less than 30 seconds of time, and should steam from a passing train or anything else obscure the view of the signal for only a few seconds the driver runs the chance of getting nothing but a glimpse of the distant signal as he passes it.

The distant signal is not pulled to CLEAR unless all the STOP signals ahead (controlled by the

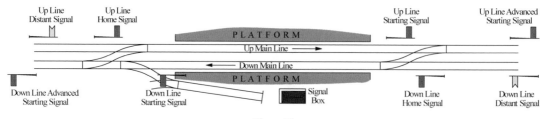

Figure 10
Main line signals at a typical station which is also a block box

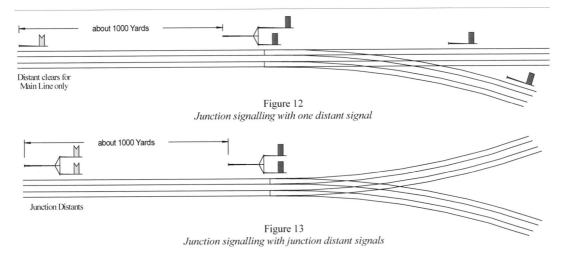

Figure 12
Junction signalling with one distant signal

Figure 13
Junction signalling with junction distant signals

same signal box as the distant signal referred to) are also at CLEAR. Hence when a driver sees the distant signal at CLEAR he takes it as an indication that he has a clear run through the station limits† into the section ahead. The distant signal being at CLEAR does not relieve the driver of the responsibility of keeping a sharp lookout for the STOP signals, and bringing his train to a stand at them if possible, should they be at DANGER, as it is quite possible for the line to have become blocked by accident after the distant signal has been passed at CLEAR, and the signalman to have put his STOP signals to DANGER before the driver reaches them.

Generally speaking, high speed running chiefly depends on the DISTANT signal, and the modern practice is to regard this signal as a SPEED INDICATOR, whereas the earliest idea was to regard the distant signal as being a REPEATER of the first STOP signal. When the distant signal is regarded as a speed indicator, it is taken to read: Arm horizontal—Reduce speed, and be prepared to bring train to a stand at the STOP signal. Arm inclined—Continue at SPEED through the station limits into the block section ahead.

Should there, however, be any reason why it is not safe for a driver to proceed at speed, quite apart from the question of the line not being clear, the modern practice is to keep the distant signal in the horizontal position. The common idea was, and to some extent still is, to regard the FIXED signals as indicating whether, from a purely block working point of view, the *signalman* can allow the train to proceed or whether he requires it to stop. If this idea is adopted a distant signal would be pulled to the inclined position although the nature of the line over which the train is required to run might require very *slow* speed for safety.

On a strip of line without junctions of any sort, should there be a very bad curve which requires to be run over at slow speed, or any other arrangement of the line permanently requiring a reduced speed it is usual to fix warning boards stating the speed allowed. Should the portion requiring a reduction of speed occur immediately after a stop signal, it would be possible to keep the distant signal permanently at danger, as a reminder of the speed restriction. This, however, is seldom done, as it would not be practicable so to warn the drivers of every portion of the line requiring reduced speed, because the slow speed portion might occur in the middle of a section, a mile or more away from a signal.

Where there is a junction, giving more than one direction for a train to run, it is not now usual to provide a separate distant signal for each direction (Fig. 12), though this practice was formerly common (Fig 13). At the junction itself there are as many stop signals as there are diverging lines. These signals act as ROUTE Indicators, telling the driver not only that he can proceed, but also the direction in which the points have been set for him to go.

At certain junctions, however, special circumstances may make it desirable to have separate distant signals for each direction. Cases of this kind might, for example, occur where high and equal speeds are permissible in each direction or where the junction is approached on a heavy

---

† Station Limits are defined as commencing at the STARTING (or ADVANCED STARTING) signal, and termination at the HOME signal of the box ahead.

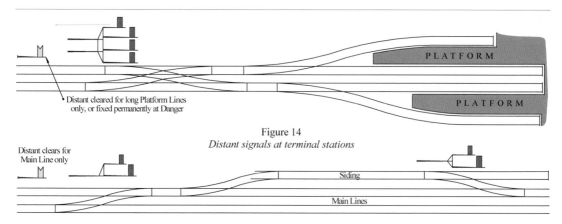

Figure 14
Distant signals at terminal stations

Figure 15
Distant signals at sidings

rising gradient, and the driver's view of the home signals is restricted by local obstructions.

In the case of a line approaching a terminal station the common practice is to provide only *one* distant signal for *all* the platform lines; that is, the distant signal is lowered if the line is clear to the buffer stops of the platform, the same signal applying to all the platforms. Should there, however, be platform lines of different lengths (Fig. 14), the distant signal should only be cleared for the long lines and *not* on any account be lowered for short ones. It is considered safer in all cases to FIX the distant signal in the horizontal position approaching terminal stations.

Where there is a facing connection into a loop line or siding (Fig. 15), no distant signal should be given for the loop line or siding, a distant signal for the main line only being provided.

COMBINING DISTANT SIGNAL AND STOP SIGNAL ON ONE POST

WHERE two signal boxes come close together, it very often happens that in attempting to carry the distant signal of the one box out to its correct distance, one of the STOP signals of the box in the rear is very near the position for the distant signal. In such cases the stop signal of the rear box is placed on the top of the post, and the distant signal placed lower down on the same post (Fig. 16). In order to allow proper clearance for the signal arms, and to ensure that the two lights at night shall be quite distinct, it is usual to place the distant signal 6 ft. below the stop signal. According to the Ministry of Transport Requirements the distant signal must *not* be placed higher up the post than the stop signal as in Fig. 17. It is necessary to equip the post in such a manner that it shall be impossible for the distant signal to be lowered until the stop signal has first been lowered, and should the stop signal be put to danger before the distant signal is returned to its normal position by the signalman working it, the mechanism on the post must put it to the horizontal position along with the stop signal (Fig. 18). Thus the top arm is said to "control" the lower or distant signal. The mechanical arrangement is termed a "SLOT," and where one signal is controlled from more than one signal box, the signal is said to be "SLOTTED." To indicate a slotted

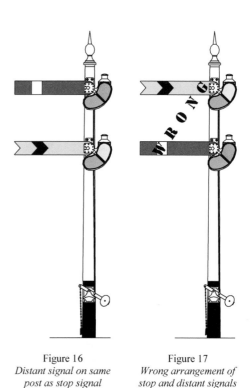

Figure 16
Distant signal on same post as stop signal

Figure 17
Wrong arrangement of stop and distant signals

signal on a plan or sketch, the usual convention is to show it with the arm in two positions, horizontal and inclined as in Fig. 19.

Sometimes where the distance between the signal boxes is very short, and when the distant signal has been combined on the same post as the stop signal of the box in the rear, it is still not sufficiently far away from its corresponding stop signal. If the rear box happens to have another stop signal in the rear of the first one mentioned, the distant signal may also be combined with that stop signal on the same post. In this case there will be two distant signals, the furthest out being termed the OUTER distant, and the other the INNER distant signal (The naming of these signals

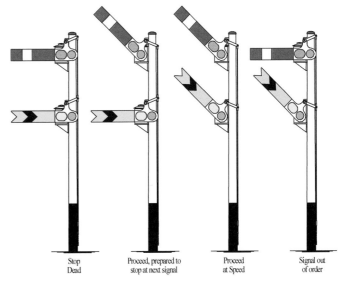

Figure 18
Aspects of semaphore signals

differs on some railways, the outer signal being sometimes termed the distant signal, and the inner one termed an AUXILIARY or REPEATER distant signal).

It might appear to be sufficient if the distant signal were carried back to the rear stop signal, without fitting a distant below the first stop signal as described, but there are one or two reasons why both signals are fitted. When a driver passes a distant signal it is usual for him to expect the next stop signal to belong to the same box as the distant signal; if, however, there is a repeater distant signal below the next stop signal it is evident that this stop signal does not belong to the same box as the distant signal. This arrangement consequently tends to avoid any ambiguity in the mind of the driver as to which signal box works the signals.

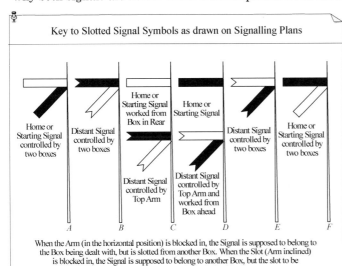

Figure 19
Method of indicating slotted signals

Placing a distant signal below the first stop signal is an advantage from a running point of view, as it is quite possible for the distant signal on the rear post to have been at CAUTION when the driver passed it, but that the signalman ahead might have cleared the line just as the driver was passing, so that when the second distant signal is seen it would be at CLEAR, allowing the driver to proceed at speed. The reverse also holds good, the OUTER DISTANT signal might have been at CLEAR when the driver passed it, but owing to some accident the signalman ahead might wish to stop the train, and have put all his signals to danger; the driver on approaching the INNER distant signal would see it at CAUTION, and receive earlier warning that he is required to bring his train to a stand at the stop signal.

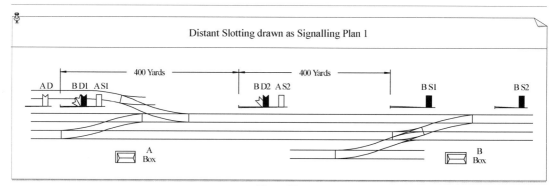

Figure 20
Diagram 1 illustrating distant slotting

Where there are two distant signals as in Fig. 20, both distant signals, BD₁ and BD₂, are controlled (or slotted) by stop signal AS₂, the outer distant signal BD₁, being slotted by stop signal AS₁ and stop signal AS₂. This is a case of double slotting, and is shown on a plan as in Fig. 21.

The necessity for this double slotting will be obvious when it is observed that without this provision it is possible for B (Fig. 20) to pull his distant signal levers before A has cleared any of his signals; assuming that B has done this, and that A wishes to lower his starting signal AS₁ to send the train to stand at AS₂, the distant signal BD₁ would move to the clear position when AS₁ was lowered, and the driver would be justified in expecting that with the distant lowered he was required to run through the section, whereas he is actually required to stop at the next signal AS₂. With the double (or back) slotting, unless both AS₁ and AS₂ are lowered, BD₁ cannot be lowered.

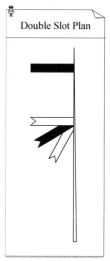

Figure 21
Symbol for double slot

In some cases, however, it is not sufficient even when the distant signal has been combined with two of the stop signals belonging to the rear box; but it is very unusual to carry the distant signal back to three stop signals of the rear box. The most common arrangement is simply to fit the distant signal on the *nearest* stop signal post (Fig. 22), and to make a rule to the effect that the signalman of the rear box shall NOT pull his distant signal to clear until the signalman of the box ahead has first pulled HIS distant signal slot lever to clear. To enable the rear signalman "A" to ascertain when the signalman ahead, "B," has pulled his slot lever (to clear), it is necessary to fit some repeating device in the rear box "A" to work in conjunction with "B's" slot lever. In addition to this it is sometimes the practice to lock up the distant signal of the box in the rear electrically, so that it is impossible for that signalman to pull it to clear until the signalman ahead has pulled his distant signal slot lever to clear, and released the lock on the distant signal of the rear box. Sometimes "B's" distant can be dispensed with by making "B" control or "slot" "A's" distant.

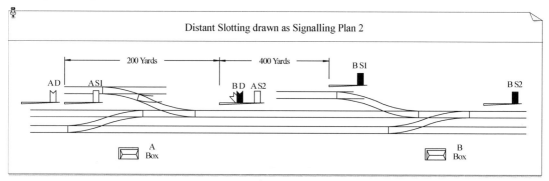

Figure 22
Diagram 2 illustrating distant slotting

In arranging distant signals for a junction, they are, as a rule, placed on brackets, but the precise arrangement of the bracket is of no importance (Fig. 23), the relative positions of the arms being the essential item. Signals for the more important lines are placed higher than signals for less important lines. Where lines are both of equal importance their corresponding distant signals are placed on the same level. Where two or more junction distant signals have to be combined with a stop signal in the rear, the stop signal must be placed over the distant signal for the most important line. Where the junction distant signals are of equal value, it is not possible to give a ruling as to which distant signal the stop signal should be placed above. If there is a distant signal on the main post, it would, as a rule, be placed above that signal; but in a case where it would improve the view of the signal were it placed to the extreme right or left, that would be done (Figs. 24, & 25).

In combining a distant signal with a set of junction stop signals, it would have

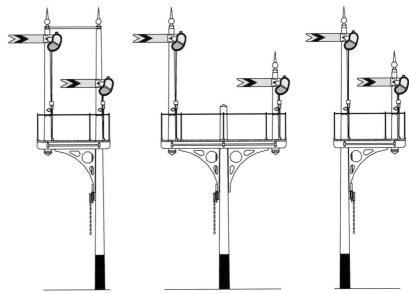

Figure 23
*Junction distant signals with various arrangements of brackets.
left hand arm—main line, right hand arm—branch line*

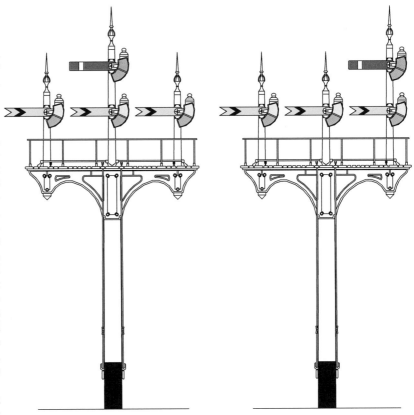

Figure 24
Possible arrangements of stop signals combined with junction distant signals

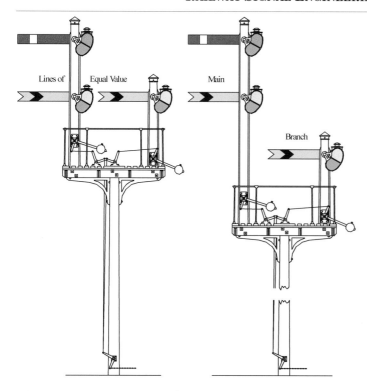

Figure 25
More possible arrangements of stop signals combined with junction distant signals

to be placed below the signal referring to its own line (Fig. 26).

Where power-operated signals are installed, what is known as a 3-position signal can be used. The indications being: (1) Arm horizontal (red light), Stop; (2) Arm inclined 45 degrees upwards (orange or yellow light), Proceed at caution; (3) Arm vertical 90 degrees upwards (green light), Proceed at speed. This signal is used in place of a combined Starting and Distant signal. This is an American idea and many of these signals are in operation in the United States and the British Colonies.

Another idea (largely adopted in India) is to place a green light 6 ft. above the distant signal light, where there is a distant signal not combined with a stop signal. The ruling then given to drivers is, "Never pass a red light unless there is a green light showing on the same post," the indications given by this arrangement at night being as in Fig. 27.

The only objection to this arrangement is the additional cost of the longer post required, the

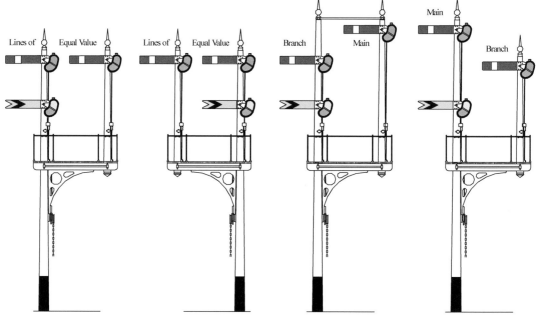

Figure 26
Distant signals combined with junction stop signals

cost of an additional lamp, and the cost of keeping it burning, etc.

The distant signal has been dispensed with on certain branch lines of the Great Western Railway (with the approval of the Ministry of Transport), and an apparatus has been installed in the driver's cab for indicating the necessary warning. A ramp is fixed in the 4 ft. space at the position where the warning is required to be given; if the driver is required to slow down preparatory to stopping at a signal at danger, an indicator in his cab shows Danger, and a whistle blows to attract his attention as his engine runs over the ramp. If the section ahead is clear, the driver is informed that he can run on at speed, by the closing of an electric circuit as his engine runs over the ramp; the current suppresses the DANGER indication and shows CLEAR instead, at the same time ringing a bell to attract his attention.

The electric circuit is controlled by the lever in the interlocking frame which would work the

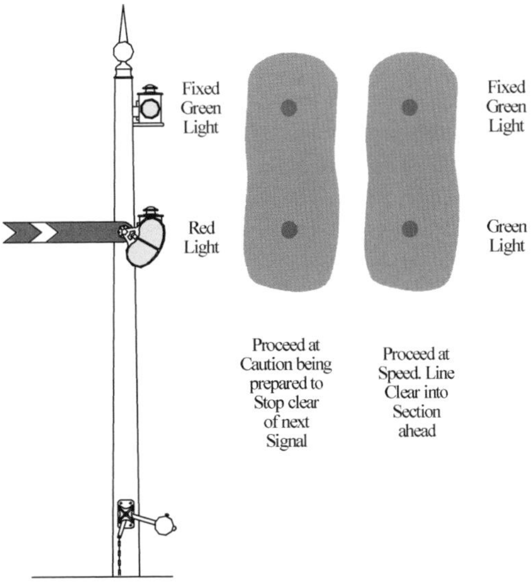

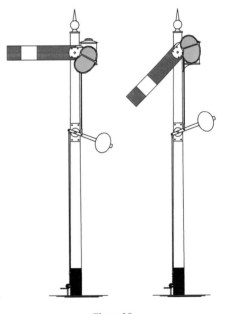

Figure 27
Indian system for distinguishing distant signal at night

distant signal if such were provided. The connection between the engine and the ramp is by means of a shoe, this shoe being attached to the engine. In running over the ramp the shoe lifts slightly, breaking a closed local circuit on the engine to give the danger indication.

HOME SIGNALS

AFTER passing the distant signal (see Figs. 10 & 11), the driver next meets the first stop signal (Fig. 28). This signal is generally termed the Home signal, it being near to the signal box or station. Being a STOP signal it has a square end to the arm, and is painted RED with a WHITE band, thus distinguishing it in daylight from the CAUTION or distant signal.

The function of the home signal is to protect a train when standing at a platform, or at a block box, from a following train, and to protect level crossing gates, point connections (cross-over roads or siding connections), or junctions.

A home signal (in fact any stop signal), in order to protect a point connection, junction, or set of level crossing gates, must be erected well clear of the exact position it is required to protect. Where a siding joins a main line the home signal must be placed in the rear of the siding connection, so that a train standing at the home signal will not interfere with the use of the siding.

A space of 6 ft. between the rails of a main line and a siding is considered to be a working clearance; anything less than 6 ft. is regarded as FOULING the main line; the point where the siding connection approaches the main line to within a distance of less than 6 ft. is said to be

Figure 28
Stop signal

the *fouling point*, and all signals must be placed in the rear of the fouling point which they protect.

At a facing junction the home signals also act as DIRECTING signals, informing the driver as to the direction in which he is being sent. Some companies treat directing signals as being in a class of their own, and do not term them "Home" signals.

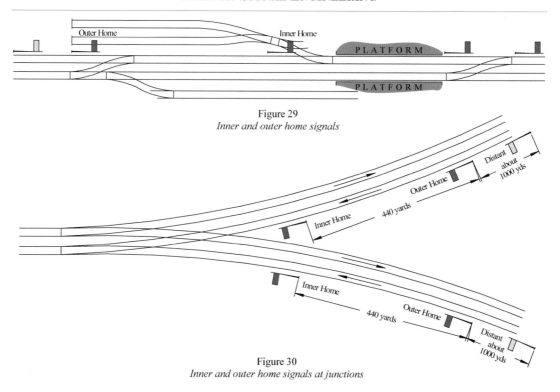

Figure 29
Inner and outer home signals

Figure 30
Inner and outer home signals at junctions

At a place where there are very many point connections it is possible for more than one home signal to be erected, and in this case the first signal encountered by the driver is called the OUTER home signal, being the furthest from the signal box; the next home signal is termed the INNER home signal (Fig. 29).

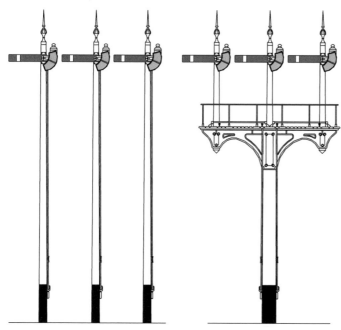

Figure 31
Two possible arrangements of junction splitting signals

It is sometimes advantageous where the inner home signals protect the fouling point of a junction to place outer home signals $1/4$ mile in the rear of the inner home signals (Fig. 30). This is to allow two trains to be accepted simultaneously on the converging lines (See "block telegraph rules" under these circumstances on p. 59). It seldom happens that the signalman can obtain a good view of both his outer home signals at a junction of this sort, and where the signals cannot be seen, they must be repeated electrically. If the train standing at the outer home signal cannot be clearly seen, or if it is more than 400 yards from the box, its presence must be indicated to the signalman in his box; usually a TRACK CIRCUIT operating an indicator is provided.

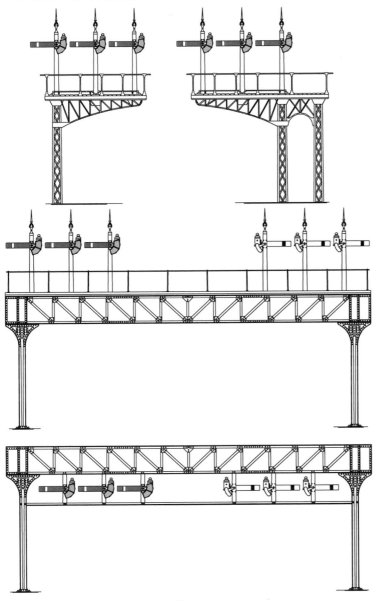

Figure 32
More possible arrangements of junction splitting signals

Where junction splitting signals are provided they must be placed either on independent posts—the posts being separate from each other—or preferably they can be on short posts carried by a bracket or signal gantry (Figs. 31 and 32). (Short posts carried by brackets or gantries in this manner are usually termed "dolls.") In the case of Shunting signals the Ministry of Transport Requirements allow the signals to be placed one below the other on the condition that the top arm reads to the extreme left, the second arm to the line next in order from the left, and so on (see Fig. 33). In arranging signal arms on a bracket or gantry, "stepping" is employed as much as possible, the heights being so arranged that the more important signals are higher than the less important ones (see Fig. 34). This "stepping" of signal arms is a great advantage in enabling a driver to pick out his signals, as should there be a row of lights all on the same level, it is by no means easy to decide at first glance to what direction any one particular signal reads.

In "stepping" signals a difference in height of about 6 ft. should be given where it is a case of main line and branch signals. Where, however, there are many signals to deal with, to give 6 ft. between each grade of signal would often make the most important signal far too high; in such a case a distance of 3 ft. makes a very good distinction; even such a slight difference in height as 18 in. can easily be picked out. Some distinction is generally made by making the signal arms of subsidiary lines shorter than main line signal arms (see Fig. 35).

INDICATING signals may be employed at stations to indicate the line or platform to which a train is going. In this device, only one arm is used for the main line from which the platform lines diverge. This arm is lowered for ALL the platform lines. Below this arm a box is fixed which shows a number or letter to the driver, thus informing him to which platform or line he is running (see Fig. 36). The numbers are about 18 in. high, illuminated at night, and can be distinguished at a distance of about 200 to 300 yards, depending, of course, very much on the state of the atmos-

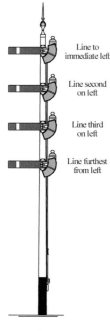

Figure 33
Shunting signals

phere. It is not desirable to employ indicating signals at fast-running junctions, as a number can be easily mistaken, and cannot be picked up so readily as the position of a light or signal arm; but at terminal stations, where there is a speed restriction of about 15 miles per hour, indicating signals are very convenient, as they save a large number of signal arms, and allow the signal to be placed over the line to which it refers.

A home signal should always be within sight of the signalman, and a train standing at it should be clearly seen from the signal box. If the signal cannot be seen it must be repeated in the signal box (this is usually done electrically); and if for any reason the train standing at the signal cannot be seen, some arrangement must be adopted to

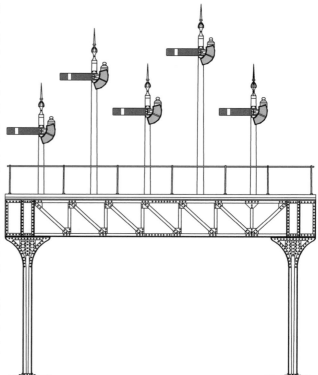

Figure 34
Signal gantry—higher signals refer to more important lines

indicate the presence of the train standing at the signal to the signalman in the box. It is very dangerous for a train to be out of sight of the signalman, as should he forget it even momentarily, he might allow another train to approach from the box in the rear, with the risk of a collision. Where a bridge intervenes and hides the view of the arm from the signalman a repeater arm is sometimes fixed lower down the post. This is commonly done to improve the view for drivers, allowing the signal to be seen first *over* and then *under* the bridge (Fig. 37).

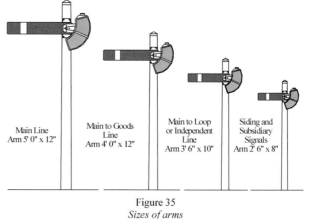

Main Line
Arm 5' 0" x 12"

Main to Goods Line
Arm 4' 0" x 12"

Main to Loop or Independent Line
Arm 3' 6" x 10"

Siding and Subsidiary Signals
Arm 2' 6" x 8"

Figure 35
Sizes of arms

REPEATER SIGNAL

IT occasionally happens, owing to curves and cuttings, a home signal must be placed in such a position that approaching drivers do not obtain a good view of it; under these conditions an additional signal may be placed further out to a position which gives a better view to the drivers. This additional signal acts as a *repeater* of the home signal, and when the repeater signal shows danger, drivers may proceed past it (as in the case of a distant signal) but must stop at the home signal should it not be at "clear" when they arrive at it. This is an undesirable arrangement and is only adopted in very unfavourable circumstances.

CALLING-ON SIGNALS

THE Calling-on signal usually takes the form of a small arm about 2 ft. long and about 6 in. wide, fixed about 6 ft. lower down on the same post as a home signal, and painted on the face, WHITE with a horizontal RED band at the top and bottom edges, the back being painted WHITE with a vertical BLACK band. The light at night should be smaller than the home signal light, a small WHITE light being exhibited for the horizontal position of the arm, and a small GREEN light when pulled off. There are several ways of treating this signal. It is now always arranged that the home and the calling-on arm *cannot both* be cleared at the same time (Fig. 38), though, in the past, some companies arranged that the calling-on arm must first be cleared before it is possible for the home arm to be cleared (Fig. 39).

The function of the calling-on signal is to allow a train to draw past the home signal at DANGER. It is used when the line is not clear far enough ahead to allow of the home arm being lowered, and it only gives the driver permission to proceed past the home signal slowly *as far as the line is clear,* stopping short of any obstruction that may be met after passing the signal.

Its chief use is to save the signalman, after stop-

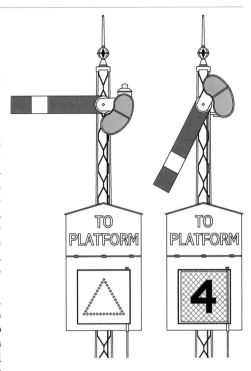

Figure 36
Route indicator signal

ping the train at the home signal, and then lowering it for the driver to proceed, from again stopping the train at the signal box to inform the driver verbally that he can proceed, but with his train under such control as to be able to stop clear of any obstruction. This signal is only provided at places where the working of traffic is such that the signalman has to stop and caution a great number of trains per day. This is the case at many stations where the platforms are sufficiently long to accommodate more than one train. If there is a train already occupying the platform, the second train should be brought to a stand, and the calling-on signal then lowered to warn the driver that he has not a clear line to run on, but that there is already a train on the line ahead of him, and he must be prepared to stop clear of it. The calling on signal is generally provided at terminal stations, even though the platforms may not be long enough to accommodate more than one train, as it is sometimes necessary to leave a vehicle at the buffer ends, in which case the calling-on signal would be given for a train running into that platform.

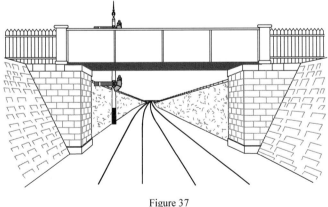

Figure 37
Signal arm repeater

The calling-on signal is also employed when backing an engine on to its train at a terminal station.

Referring back to the different arrangements of working the calling-on signal by the various companies, and taking the arrangement in which the home and the calling-on signal cannot be lowered at one and the same time, it is assumed that each arm, when lowered, conveys a different meaning. The home arm indicates "PROCEED, LINE IS CLEAR;" the calling-on arm indicates "PROCEED AT CAUTION, line is already OCCUPIED." The home arm in each case acts as the STOP

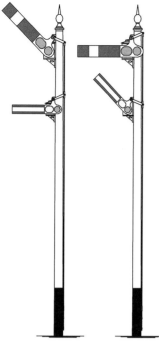

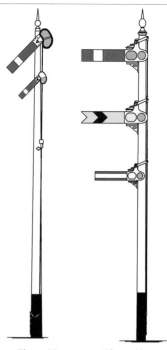

signal, while the small calling-on signal gives permission to move past it slowly. With night working, the large RED home light is first seen plainly indicating STOP, and when the train has come quite to rest, the small GREEN light gives the permission to pass. When the home arm is lowered for a clear line there is still the small WHITE light of the calling-on signal which is *not* required, hence some companies (as mentioned before) do not show any light for the horizontal position.

Where only one signal is given at one time, should it be found necessary to provide a calling-on signal on a post which has a distant signal combined with the home signal, the calling on signal is placed *below* the distant signal (FIG. 40).

With the arrangement where the calling-on signal is lowered before the home signal, the *whole* combination, instead of the individual arms, is taken as giving the indications to the drivers. Thus when the line is clear right through, both the small calling-on and the large home signals are lowered; at night this, of course, shows two green lights for clear. With this arrangement should it be found necessary to provide a

Figure 38
Calling-on signals

Figure 39
Small arm calling on signal

Figure 40
Calling-on signal with home and distant arms

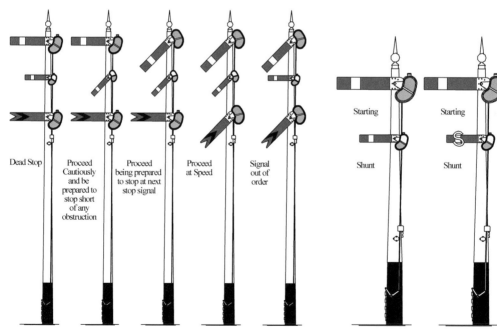

Figure 41
Former N.E. Rly. arrangements of calling-on arms

Figure 42
Shunting signals

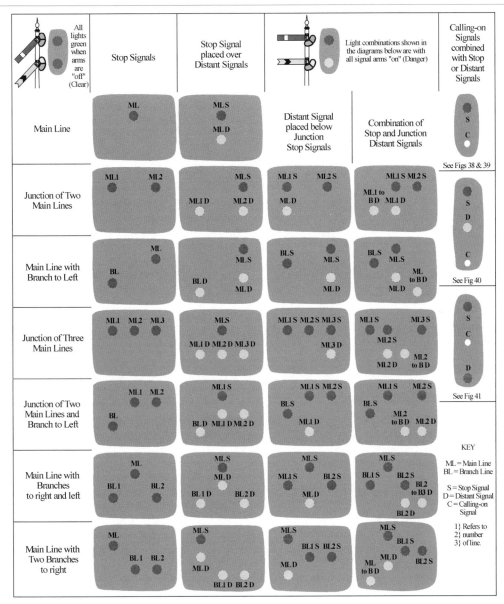

Figure 43
Groups of combined signals as seen at night

calling on signal on the same post as a home and distant signal, the calling-on signal is placed *between* the home and distant signals (Fig. 41).

STARTING SIGNALS

THE starting signal is a STOP signal, and is usually placed at the end of a platform, the idea being that should a train have been allowed to pass the home signal in order to stand at the platform, it will hold the train at the station until the signalman has permission for it to proceed into the next section. Should there be another signal ahead of it, the starting signal loses its importance and then simply becomes an *intermediate* STOP signal.

ADVANCED STARTING SIGNALS

THE advanced starting signal is also a STOP signal, and always controls the entrance of a train into the section ahead. A train having passed the advanced starting signal (except for shunting purposes) is out of the control of the signalman working that signal, and it comes under the direction of the signalman ahead.

The advanced starting signal (sometimes referred to as the ADVANCE signal) is usually placed beyond the furthest out set of point connections. The distance it is out from these points depends largely on the nature of the traffic working, but, if possible, sufficient head room is given for the longest train using the sidings to draw up to the signal with the rear vehicle clear of the points, to allow of the train being backed into the siding if required. Sometimes only an engine length is sufficient at one place, whereas 400 yards may be required at another station. If, as sometimes happens, an engine or train standing at the signal cannot be seen, or if the signal is more than 400 yards from the signal box, some device, *e.g.*, track circuit (see p. 47), must be adopted to indicate the presence of the train to the signalman in the signal box.

SHUNTING SIGNALS

THE shunting signal is a small arm fixed lower down on the same post as a starting or advanced starting signal, and its function is to give permission for a train to pass the Stop signal, when that signal is at danger, for SHUNTING purposes only (see Fig. 42). This signal is used when it is not possible to place the starting (or advanced starting) signal sufficiently far out from siding points to enable a train to draw up to the latter with the rear vehicle clear of the points. The lowering of the shunting signal authorises the driver to pass the starting or advanced starting signal at danger into the section ahead far enough to allow the last vehicle to clear the siding points, but on no account must a driver go on to the next signal box until the starting or advanced starting signal has been lowered. At night, either no light is shown or a red one is given for the horizontal position, with green for the inclined position. Where there is a distant signal on the post with the starting signal, the shunt signal is treated in the same manner as the calling-on signal.

The foregoing are the signals commonly used for main line working. At very complicated junctions and stations the names used, *viz.*, Outer Distant, Inner Distant, Outer Home, Inner Home, Starting and Advanced Starting, are not sufficient to classify all the signals employed. In such cases special names have to be assigned to the signals to meet the situation; but the modern tendency is to reduce the number of running signals to the minimum for the convenience of the drivers.

On some railways, where there are duplicate or loop lines, the signals for these lines are distinguished from main line signals—for daylight working—by having rings attached to the arms.

Fig. 43 shows the appearance at night of the various signals which are commonly encountered.

COLOUR LIGHT SIGNALS

COLOUR light signals, which have recently come into favour, are found in several forms, all of which exhibit aspects consisting of powerful Red, Yellow or Green lights (Fig 44). The source of light is electricity, the change in colour being effected in some cases by having a mechanically or electrically operated spectacle with coloured glasses in front of a clear glass lens. In other cases separate coloured lenses are provided for each colour and the change in aspect is secured by switching on the lamp behind the appropriate coloured lens. Colour light signals have been designed which are plainly visible from a distance of over 1,000 yards in bright sunlight.

The indications of Colour light signals have the same meaning as the 3-position signal mentioned on p. 14, *i.e.*, RED—stop; YELLOW—proceed with caution prepared to stop at next signal which is at "stop;" GREEN—proceed, next signal is at Yellow or Green.

Figure 44
Colour light signal

The advantages which have led to Colour light signals being increasingly adopted both for main running lines and terminal stations are:

- Simplicity of aspect, day and night indications being the same. The small size of the signals in many cases enables them to be placed in closer relationship to the lines to which they apply than is possible with semaphore signals.

- Increased visibility in dark or foggy weather; under which conditions semaphore signals are at a disadvantage. Their visibility in fog is such that some Railway Companies do not provide fogmen for them (see below).

- Simplicity of construction and consequent ease of maintenance as compared with power worked semaphores.

FOG SIGNALLING

IN foggy weather, or during falling snow, when the drivers cannot see the signals clearly, it is the general practice to station ground signalmen with flags or lamps and detonators at the DISTANT signals. They are instructed to maintain two detonators 10 yards apart some distance in the rear of the signal which they have to repeat, so long as that signal is in the danger position; when the signal moves to clear the detonators should be removed. In addition to placing the detonators on the rails for the on-coming train to explode, each fogman must have hand signals for showing RED or YELLOW and GREEN. The purpose of the detonators is to draw the driver's attention to the hand signals; if the signal is cleared, whether the detonators are removed or not, the fogman must exhibit a green hand signal to the driver. If the driver explodes a detonator and does not receive a hand signal he must regard it as a DANGER signal and be prepared to stop short of any obstruction.

In the event of the signalman requiring a train to stop during the time between calling out the fogmen and their arrival at their posts, he must endeavour to place two detonators on the rails opposite his signal box.

In many cases this is difficult to carry out, and some mechanical device worked by a lever in the signal box is desirable for placing the detonators on the rails, especially where the signalman has to cross a set of running lines to carry out the rule.

A detonator-placer might be in operation, working permanently in connection with the home signal lever, so that should a train over-run the home signal at any time an audible warning would be given. Of course the signalman would have to replace the exploded detonators in the machine after a train had over-run the signal, but in ordinary working that would be *very* seldom. Where this device is used it is necessary that the detonators be changed at regular intervals to ensure their being in good condition.

Numerous devices have been, and are being, invented to dispense with the services of fog signalmen. They usually take the form of automatic magazine devices controlled from the signal box, or some form of "CAB SIGNALLING," where an indication is given to the driver in his cab.

Generally speaking, no form of fog signalling device is likely to be successful which requires to be worked by the distant signal *wire;* where distant signals are about 1,000 yards from the box, and occasionally work more than one slotted signal, that alone is sufficient duty for that wire to perform satisfactorily without extra weight being added.

No device which depends on pendulums or triggers on the engine striking a projection in the 4 ft. space (or by the side of the line) is likely to be very successful, as with express speeds the impact is very severe and renders the fittings costly to keep in efficient working order.

The following features are necessary for a good working fog signalling device of the CAB SIGNALLING TYPE:

1. Distinctive indication for both CLEAR and DANGER (the absence of an indication is not sufficient for CLEAR).

2. The apparatus should be constantly in service during all weathers, both foggy and clear.

3. It should work in conjunction with the distant signal, but should not add any additional work to the lever of that signal.

4. Any portions on the engine which move when passing over the apparatus should have the minimum amount of INERTIA.

It is almost impossible to design a purely mechanical device to fulfil all these requirements. The employment of electricity appears to be necessary for the operation of a successful device. There are a few types of apparatus employing electricity, and as a rule they employ skids on the engine, which make an electric contact with a ramped bar placed in the 4 ft. way, as mentioned on p. 15.

Another distinct type of automatic fog signalling machine employs a magazine containing cartridges, which are exploded electrically on the train running over a treadle when the signal is at danger.

There is a very useful type of fog signalling machine on the market (Messrs. Clayton's) with a magazine which prevents the fogmen from having to place the detonators on the line by hand. Where there are more than two lines, and the fogman is required to cross over a running line to place the detonators, the fixing of a magazine machine saves the fogman from having to cross the lines to do his work. The machines are worked with rodding from small levers fixed by the side of the line.

Instead of two separate detonators, where machines are used, it is usual to employ a double detonator, which gives only one report on exploding ; this saves the fixing of two machines.

A machine is also on the market for saving one of the two detonators, except in the event of the first one proving defective; the sole purpose of using two detonators is to safeguard against one being defective. With this device the detonator which would be exploded last is fixed to a strip of metal, the strip of metal in turn is connected to a small shaft, while on the opposite end of the shaft a shield is fitted. The apparatus is fixed by the side of the rail with the shield opposite the first detonator, the first one being placed on the rail by hand in the usual manner. If the first one explodes as the train runs over it, the force of the explosion acts on the shield and causes the shaft to turn and take off the second detonator before the train reaches it.

It is usual to provide small huts for the fogmen, and occasionally, where the clearances are narrow, fog-pits are sunk for the men. Where very dense fogs are experienced, it is necessary to fix some type of indicator to repeat the action of the signal, otherwise it is not possible for the fogman to ascertain the position of the signal arm he is intended to indicate. This may be done by the signal wire or electrically. It is common to place short arms low down on the post working in conjunction with the distant arm for this purpose.

SIGNALS FOR CONTROLLING SHUNTING MOVEMENTS, ETC., BETWEEN SIDINGS AND RUNNING LINES

1. *Shunting from a Siding over a Running Line in the Wrong Direction on to a Line in the Right Direction.*

An ordinary ground disc signal or a small semaphore signal is generally used for this movement; some companies, however, employ a semaphore signal of distinctive shape as in Figs. 45 and 46.

2. *Shunting from a Siding on to a Running Line in the Wrong Direction.*

Signals of this type must only be provided where steps are taken to ensure that, in the event of a misunderstanding between the driver and the signalman, or shunter, the train will not proceed on the running line in the WRONG direction.

The ordinary method of ensuring this is to compel the signalman to set either the trailing points to run it to a siding, or the cross-over road points to run it on to a right line (should such exist) *before* the WRONG LINE signal can be lowered. If it is not possible to turn the train into a siding, or on to the right line, or if it would seriously hamper traffic to set the points for the train to go on to the right line, then a "Limit of Shunt" Indicator should be placed on the line used in the wrong direction, at a position beyond which shunting must not proceed, and it may also be necessary to fix a home

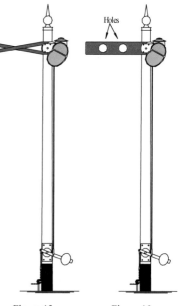

Figure 45
Wrong line signal

Figure 46
Right line via wrong line signal

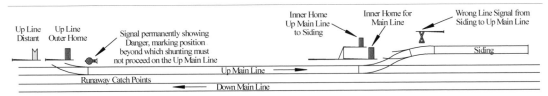

Figure 47
Wrong line signalling

signal outside the indicator to hold back trains whilst shunting is in operation (Fig. 47).

3. *Shunting to and from Sidings and Shunting Neck.*

A signal is employed on some railways to indicate to a shunter when it is safe to shunt in and out of a group of sidings to the shunting neck; as this signal allows the train to shunt in both directions it is called a BOTHWAYS signal (Fig. 48). Both sides of the arm are painted RED with white stripe, and it has a lamp and glasses to show red and green in *both* directions. To distinguish it by day a circle is attached to the arm, which is a small one. When the signal is at DANGER the shunter must keep his train clear of the points giving access to the sidings, but so long as the signal is at CLEAR he is assured that the signalman cannot move the points to let a train into the sidings, or move the points in any way to interfere with shunting operations (see Fig. 49).

Figure 48
Bothways or loop siding signal

4. *Shunting from a Main Line to Sidings, from Sidings to Main Line, and from one Main Line to another by means of Cross-over roads.*

Signals for these movements are generally Disc signals, or very small semaphore signals fixed low down (Figs. 50, 51, 52); whether they are disc signals or small semaphore signals they are usually termed GROUND signals (sometimes referred to as *dollies* or *dummies*). It sometimes happens that the ground signals cannot be seen by the driver owing to their being too low down; in this case it is usual to raise them on posts, or to fix small semaphore signals of shunt type on posts about 12 ft. high (Fig. 53). If the signal has to be placed in a space of less than 10 ft. between the running lines, it is necessary that it should *not* be more than 3 ft. above rail level to meet the Ministry of Transport requirements.

When a Disc signal is at "DANGER" a RED or YELLOW disc faces the driver, and when the signal indicates ALL RIGHT the disc is either turned to one side, with the edge (or a small green disc) facing the driver, or the red or yellow disc is lowered to a horizontal position showing only its edge.

For night working the light normally shown may be either YELLOW or RED, the latter being used for signals which must not be passed at Danger without special permission; for all others YELLOW is used. Both types show GREEN for "All right."

There is great objection to a large number of BIG RED lights facing a driver when he is running through a station yard at night. When close to the signals it is quite easy to see that they are ground disc signals, but at a distance it is difficult to decide which are running signals and which are siding signals, especially if the line is on a curve and the disc signals not well placed. For this

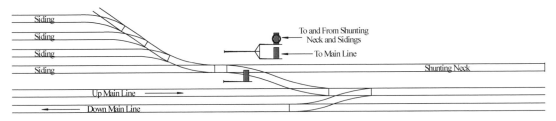

Figure 49
Bothways signalling

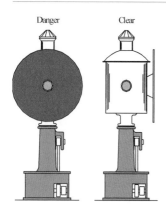

Danger   Clear

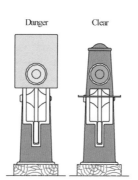

Danger   Clear

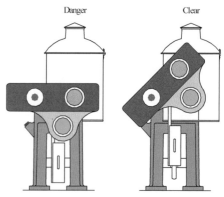

Danger   Clear

Figure 50
Ground disc signal rotating type

Figure 51
*Ground disc signal
flap type*

Figure 52
Ground semaphore signal

reason it is becoming increasingly the practice to pull off the Disc or Ground signals applicable to a certain route before pulling off the semaphore applying to that route.

Where shunting signals apply for trains entering the Block Section ahead—as is the case of a signal for leaving a siding when there is no starting signal ahead—a semaphore arm of small size is preferred to a ground disc signal.

Paragraphs 2, 3, 4 and 5 of the MINISTRY OF TRANSPORT REQUIREMENTS refer to SIGNALS, and read as follows:

SIGNALS

2. In the interests of economy, and to avoid confusion, the number of signals provided, and their height, should be limited to what is actually necessary for safety and traffic purposes.

Up and down Distant signals for each block post, to be provided on all running lines which have two-position Stop signals. One Distant signal in each direction, with the necessary control from each signal box, is sufficient for a number of block posts closely grouped together, unless there are good reasons to the contrary.

At diverging junctions, one Distant signal only should be provided, worked for the junction line over which the highest speed is permissible, unless in exceptional circumstances more are essential.

Where special circumstances, e.g., permanent speed restriction, justify the adoption of an unworked signal, it should be secured in the warning position and not coupled up or duplicated for directing purposes.

The interval between a Distant signal and the first Stop signal to which it applies must be such that a train in proximity to the former, and moving at the highest authorised speed, can be stopped before passing the latter.

Stop signals to be provided for each up and down line at all block posts. At diverging junctions, a separate running signal will be necessary for each direction of movement. Where outer and inner Stop signals are provided on the approach to diverging junctions, it will not be necessary to give the full route indication at the former which is given at the latter.

At stations with a number of diverging lines, one signal with indicating apparatus for each approach line should, as a general rule, be provided instead of separate signals. When, however, there are through fast lines, a separate arm should be provided for each. On passenger lines, all connections within yard limits to be under the protection of Distant and Stop signals.

3. All signals, as a rule, to be immediately on the left of, or vertically over, the line to which they apply. At diverging junctions, bracket signals are preferred to signals carried on separate posts,

Short Arm
Small Light

Figure 53
*Semaphore
siding signal*

unless there are reasons to the contrary.

In the case of shunting signals, where more than one are necessary, direction may be indicated by carrying them vertically one below the other, in which case the top signal will apply to the line on the extreme left, the second signal to the line next in order from the left, and so on.

Semaphore Distant signals to be distinguished from Stop signals during daylight by Yellow-coloured arms, with notches cut out of the ends. They must be placed below, and be controlled by, Stop signals, if these are carried on the same post and applicable to the same direction. A Distant signal placed under a Stop signal of the box in rear must unless the circumstances are exceptional, be repeated under all Stop signals in advance of that signal which are worked from that box, with the necessary additional control by such signals.

Signals for shunting movements should be readily distinguishable from running signals. They should therefore be placed as close to the ground as the circumstances permit, and should be of the miniature arm or other approved type, with small lights.

The facing side of the arms of all semaphores (including miniature), and the face of disc signals to be painted to accord with the colour of the light exhibited in the Danger or Caution position.

A special type of shunting signal for wrong line movement is not considered necessary. In such cases where it is not possible to turn the movement in the right direction on to a running line by reversing a cross-over or on to a siding by reversing the points, an indication, visible by night and day, of the limit of such movement will meet the case.

With semaphore signalling, indications for "calling-on" movements to be given by a small arm carried under the relative Stop arm. By night, a White light to be shown in the normal position, and the light authorising the "calling-on" movement to be Green. "Calling-on" signals to be used only for the specific purpose of indicating to the driver, either that the line between the "calling-on" signal and the next Stop signal (or buffer stop, when there is no Stop signal in advance) is occupied, or that he is required to stop for instructions at the signal box ahead. The "calling-on" signal should not therefore be capable of being worked at the same time as the relative Stop signal.

Stop signals when working automatically under the "Stop and Proceed" regulation should be distinguished by the letter "A." In the case of controlled signals when working automatically the letter "A" should be visible both by night and by day, and should be obscured when the "Stop and Proceed" regulation does not apply.

For two-position semaphore signals, the arm indication to be horizontal for Danger, and 45 degrees for Clear.

Light signals of an approved type should be used, in lieu of semaphores, for three or more aspect signalling. They may also be used for two aspect signalling.

Signals to be so designed as to give a Danger indication in the event of any failure of the mechanism which operates them.

4. *Front lights of all running signals to be Red for Danger, Yellow for Caution (including the warning position of Distant signals), and Green for Clear. These colours, in each case, to be within the approved standards. White to be used for the back lights of signals.*

With semaphore signalling, back lights, visible only when signals are at Danger, and no larger than actually necessary, to be provided when the signalman cannot see the front light of any signal which he works, the arm of which is visible by day. Back lights should be provided for all ground signals.

For two-position shunting signals, the normal light indication may be either Red or Yellow. The Red light to be used only when it is necessary to indicate that the signal is not to be passed without special permission, unless it is in the Clear position. In other cases Yellow to be used. In the case of shunt ahead signals, etc., carried under running arms, the lights used to be the same as those for "calling-on" signals, with lettering as necessary.

The arms of all Stop signals, and the Danger or "on" aspect exhibited by all light signals, which cannot readily be seen by the signalman, and the arms of all Distant signals, to be repeated in the signal-box from which they are worked. It is desirable that the lights of all semaphore Stop and Distant signals upon important lines with high speed traffic, should be repeated, unless either the front or back lights can readily be seen from an adjacent signal-box.

To prevent confusion with signal lights, the use of coloured lamps for engine head lights or other purposes should be avoided. Red to be used for tail lights.

A Red light to be used by night to define the position of buffer stops at the termination of platform arrival lines.

5. *All worked signals, except where necessary at level crossings, should, as a rule, be dispensed with under the following conditions:*

On single lines—

a) *At all stations and siding connections upon a line worked by one engine or motor vehicle (or two or more such engines or vehicles coupled together) carrying a staff, when all points are locked by such staff.*

b) *At any intermediate siding connection upon a line worked under the train staff and ticket system, or under the electric token system, where the points are locked by the train staff or electric token.*

c) *In special conditions, signals may also be dispensed with at crossing loops upon a line worked under the train staff and ticket system, if the loop points are locked by the relevant staff.*

d) *At intermediate stations which are not token stations, upon a line worked under an electric token system; sidings, if any, being locked as in (b).*

On double lines—

a) *At an intermediate siding connection, either where "lock and block" or other similar approved apparatus is in use, or where the points are mechanically or electrically controlled from one or both of the adjacent signal-boxes, and the relative running signals suitably interlocked.*

b) *At stations which are not block posts, where there are no connections.*

3

CONSTRUCTIONAL DETAILS OF SIGNALS

SIGNAL POSTS

SIGNAL posts are commonly made of wood or of iron lattice-work. Some posts are made of hollow tubing; these, however, are not common, except in connection with power-worked signals, or light signals; also recently some have been made in reinforced concrete.

There is very little to choose between a wood post and a lattice iron one. For first cost the iron post is dearer than the wooden one, but its life is correspondingly greater. The precise life of a post depends largely on the soil and atmosphere surrounding the post.

Taking an average, a good wood post might in this country last from 25 to 30 years, whilst under favourable circumstances an iron post might last from 30 to 40 years, or even longer. A wood post is heavier to handle than a lattice iron one, and offers a larger area to the wind. It is, however, slightly more conspicuous than an iron one, and on that account is preferred by some companies.

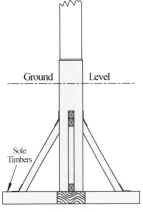

Figure 54
Base of wood post

A wood post is usually made of red pine or other similar timber, which must be sound, free from sap and straight grown. The timber is trimmed from the log to be about 10 in. or 12 in. square at ground level (depending on the height of the signal) to about 6 in. square at the top. The end of the post to be let into the ground is fixed to 12 in. by 6 in. cross-timbers with struts and through bolts; and with the depth of the signal in the ground varying with the height (Fig. 54).

The length of the sole cross-timbers also depends on the height of the signal, a good average being as follows:

Height of Signal	Base in Ground	Length of Sole Timbers
	About	About
20 ft.	4 ft. 6 in.	4 ft. 6 in.
30 ft.	5 ft. 0 in.	5 ft. 0 in.
40 ft.	6 ft. 0 in.	6 ft. 0 in.

The whole of the base, and up to 2 ft. above ground level, should be dressed with hot gas tar. The top of the post requires to be fitted with a cap or pinnacle to keep out the wet. All fittings on the post are either fastened with through bolts or with coach screws. One of the disadvantages of the wood post is the tendency for rot to set up at the coach screws and bolts, also at a point just above ground level. To prevent this as much as possible, the post should be very thoroughly painted before and after erection.

Iron posts are usually built up from angle-iron and flats, a common size of angle-iron being 1¹/₄ in. by 1¹/₄ in. by ¹/₄ in., the flats 1¹/₄ in. by ³/₁₆ in., and the rivets ⁵/₁₆ in. snap heads. For posts up to 25 ft. high single bracing (see Fig. 55) is used, but for all posts above that height double bracing is resorted to, with a double set of angles up to about 20 to 25 ft. Where brackets have to be attached to the posts double bracing is always employed, and double angles up to the brackets. The ends of the angles are fastened to a cast-iron base by hoops which are forced on and they hold by friction, the post and base having a slight taper. Where the rings are forced on the angles are strengthened by 1¹/₁₆ in. by 1¹/₁₆ in. square iron (see Fig. 56). The top of the post is either fitted

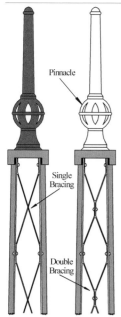

Figure 55
Lattice post with pinnacle

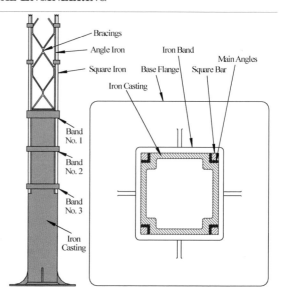

Figure 56
Base of iron lattice post

with a cap or a pinnacle in order to give it a good appearance. The cast-iron base is set on timbers at the bottom of the post hole in order to give it a good even bedding. All fittings on a lattice iron post are, as a rule, held on by friction, the parts being clamped on with straps and bolts. Once the post is in service there is little or no trouble experienced from the fittings slipping if they have been properly tightened up to commence with. The post must be carefully painted to preserve it from rust.

Unless the joints are well coated with paint sufficient to keep the wet out they are usually attacked by the rust forcing apart the bracing from the angle irons.

Signal posts are generally made in standard heights, the height being measured from the rail level to the centre of the arm. Where posts are placed on the side of an embankment or cutting the height is taken from a point above ground level which would be equivalent to rail level under ordinary conditions. The usual standard heights are: 15 ft. (chiefly siding signals, or where the signal has to be seen below a bridge), 20 ft., 25 ft., 30 ft., 35 ft., 40 ft., 45 ft. and 50 ft., the average being 30 ft. Above 45 ft. is exceptional.

The standard height for a main post carrying a double bracket is 20 ft. from rail level to the top of the bracket, adding some multiple of 5 ft. if a greater height is required. All posts higher than 25 ft. should be guyed to give stability. It is essential that the guy wires be fixed so as not to cause any obstruction to persons who have to pass the signal. If the signal is placed in such a position that men are constantly passing it, the guy wires must not be fixed direct to the usual anchor bolts in the ground, but should be fixed to stay posts instead; the reason for this being that a man in the dark can keep clear of a post more easily than a wire.

BRACKET SIGNALS

MANY signals are carried on brackets fixed to a main post. For wood posts these brackets are usually made of cast iron, or if the bracket has to be made very long, or carry a great weight, a bracket of the lattice iron pattern is built up. For iron posts a plain bracket of angle and flat iron is generally employed (Fig. 57). The length of bracket varies with the length of arm to be carried, or cleared, as the case may be. The usual lengths are 5 ft., 6 ft. and for full-sized main line arms 6 ft. 6 in. Sometimes the bracket is made longer to clear some obstruction hiding the signal, in which case a special bracket is made up to suit the circumstances. Brackets are constructed to carry one short post or doll, double brackets to carry two dolls—one on either side of the main post—and triple brackets to carry three dolls—one on the centre line of the main post and one on either side of it. To the top of the bracket a landing or gangway is fixed for the convenience of mechanics, signal lighters and cleaners. All brackets should have a uniform method of fixing the dolls, so that the ends of the dolls can be made to standard pattern. The dolls are made either of wood or iron, but generally of the same material as the main post. It is usual to arrange for the dolls being made to the following standard heights:

For one arm	5 ft.	8 ft.	11 ft.	14 ft.	17 ft.
For two arms, home and distant			11 ft.	14 ft.	17 ft.

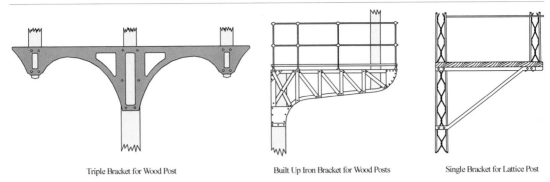

Triple Bracket for Wood Post   Built Up Iron Bracket for Wood Posts   Single Bracket for Lattice Post

Figure 57
Signal brackets

This allows for stepping the heights of the signal arms 3 ft. per step. With dolls above 11 ft. they should be stayed across to the main post or to any other dolls in the group, and the bracket itself should be guyed to anchor bolts in the ground or to stay posts.

SIGNAL BRIDGES

IT is frequently necessary to erect signal bridges to carry a group of signals—which may be too many to be accommodated on brackets—and this arrangement has also the advantage of placing the signals near the lines to which they apply.

Any sort of bridge can be utilised, but it is preferable to have a bridge built specially for the place and purpose. One of the best types of signal bridge is a common lattice girder fastened to strong upright posts. For bridges spanning not more than four sets of lines (about 50 ft. long) a girder 3 ft. deep is suitable (Fig. 58). The sizes of the angles, etc., vary with the span, but for a bridge up to about 50 ft. span the dimensions would be as follows: Main angles, 3 in. by 3 in. by $1/2$ in.; vertical angles and cross bearers, $2^{1}/_{2}$ in. by $2^{1}/_{2}$ in. by $3/_{8}$ in.; flat braces, $2^{1}/_{2}$ in. by $1/_{2}$ in.; rivets $5/_{8}$ in.

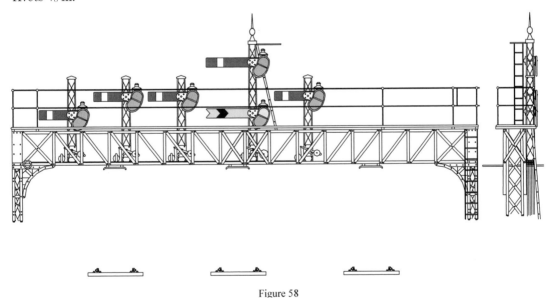

Figure 58
Girder signal bridge

The posts are built of angles and flats, the angles being about $2^{1}/_{2}$ in. by $2^{1}/_{2}$ in. by $3/_{8}$ in., and the flats $2^{1}/_{2}$ in. by $1/_{2}$ in. The base is made of cast iron with the angles of the post wedged firmly with three bands. The angles at the base should be strengthened with outer angles 3 in. by 3 in. by $1/_{2}$ in. (see Fig. 59).

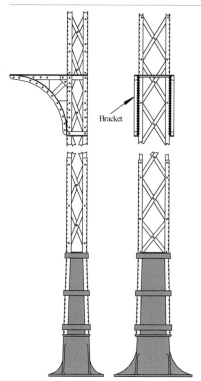

The posts of a standard pattern bridge should not be wider than 12 in. to enable them to be set down in a space of 10 ft. or 11 ft. between the running lines to clear the structure gauge; the side parallel to the lines being not less than 18 in. A bracket is generally fitted to the top for the purpose of supporting the girder, but in some designs the bottom angles of the girder are bent into the main posts, and bolted or riveted to them. The bridges are fitted with landings and hand-rails, similar in design to those used for bracket signals. If wood dolls are to be erected, sockets should be made and fastened to the girder to receive the ends of the dolls, these sockets being the same as for the standard pattern bracket doll. The ends of iron lattice dolls are fitted either with a plate for bolting on to the girder, or made with ends to receive bolts, in which case a plate is riveted to the girder to receive the dolls.

LADDERS

LADDERS have to be fitted to all signal bridges and bracket signals to enable workmen to get to the landings. All posts and dolls higher than 6 ft. should be fitted with ladders up to

Figure 59
Leg and base for signal bridge

the arms (Fig. 60), with landings about 3 ft. below the arms to enable the lamps and glasses to be cleaned easily. Sometimes, instead of landings at the top of the post, the ladder is carried to the top with a ring or bow on it to prevent men falling when reaching round to the opposite face of the signal for cleaning, oiling, or repair purposes.

With ordinary main posts some companies used to provide winding gear for letting down the lamps for trimming, and in this case do not always provide ladders.

There is some diversity of opinion as to which side of a post the ladder should be fixed, but the usual practice is to place the ladder so that a man in climbing down would FACE an approaching train. This would, as a rule, place the ladder at the back of the post. There is, however, some objection to this, as when a man is at the back of the post he cannot see if the face side is clean and in good condition unless he takes the trouble to lean round. If the ladder is placed on the face side, a man can easily see the condition of the spectacle glasses, and, without having to reach

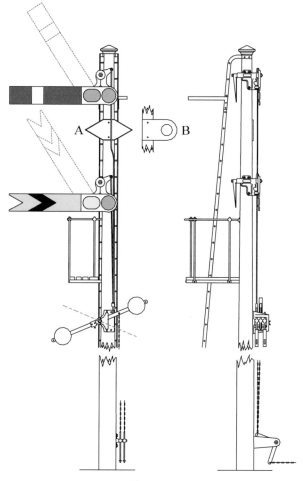

Figure 60
British Standard upper quadrant signal, ladder and fittings

round the post, can make an examination of the signal.

The usual size for ladders for main posts is 12 in. wide, and the pitch of the rungs 11¾ in. For short posts and dolls the ladders usually are about 10 in. wide and about 9½ in. pitch for the rungs. The sides of the ladder are made of 1½ in. flats, and the rungs of ¾ in. round iron.

SIGNAL ARMS AND SPECTACLES

SIGNAL arms used to be made of wood (generally cedar or mahogany, occasionally of pine). The new British Standard arm is, however, of enamelled steel. The tip of a wood arm is about ¾ in. thick, and the head about 1 in. thick. The thickness of the steel arm is about ³/₃₂ in., and the corrugations shown are to impart sufficient stiffness to prevent its being bent in a gale (Fig. 61).

The lengths of wooden signal arms generally employed were as follows:

Main Line	Main Line to Goods Line	Main Line to Siding	Sidings
5 ft. 0 in. × 12 in.	4 ft. 0 in. × 12 in.	3 ft. 0 in. × 9 in.	2 ft. 6 in × 8 in.
or	or	or	or
4 ft. 6 in. × 10 in.	3 ft. 6 in. × 10 in.	2 ft. 0 in. × 6 in.	2 ft. 0 in. × 6 in.

Figure 61
Section of steel arm

These dimensions are from the centre of the signal post to the tip of the arm. Some companies had the sizes of wooden arms made proportional to the height of the signal post, a high post having a longer and broader arm than a short post. As a rule the arms on brackets were made shorter than the arms on single posts, in order to keep the brackets as short as possible.

A common method of fitting the arm to the post is to bolt it to a cast-iron plate, this plate being keyed on to a spindle. The spindle turns in a cast-iron bearing which is fixed to the signal post.

Generally the spindle was fitted at the head of the arm in such a way that if an arm was 5 ft. overall, the tip of the arm would swing with a radius of 4 ft. 6 in. This necessitated the spindle being fitted through the centre of the signal post (Fig. 62). If the spindle was fitted with its bearing just clear of the post, then the spindle would be so placed on the arm that the tip of the arm would swing with a radius of about 3 ft. 10 in. (see Fig. 63). One or two companies used to fix the spindle approximately in the centre of the arm, in a position slightly *above* the centre line, so that the arm balanced itself (see Fig. 64). This type of arm requires a long bracketed bearing to carry the spindle at a distance of about 2 ft. from the edge of the post.

If the spindle is fitted at the head of the arm, it is, with lower quadrant arms, necessary to counterbalance the arm, so that in the event of the connections between the arm and the lever breaking, the arm will not droop, and should it be in the CLEAR position when the failure takes place, it must instantly go to the DANGER position (see Ministry of Transport Requirements, p. 27). This is effected by fitting the SPECTACLE on to the same casting to which the spindle is keyed, and making the spectacle sufficiently heavy to counterbalance the arm. In several

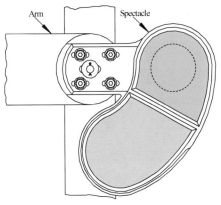

Figure 62
Spindle through centre of post

types of signals the spectacle is fitted separately to the spindle. In designing the spectacles of lower quadrant signals, for counterbalancing purposes, it is necessary to give a good margin of safety in favour of the spectacle, as it is possible that although the spectacle easily puts the arm to DANGER when the signal is first erected, the coatings of heavy paint received by the arm during years of service may eventually make it too heavy for the spectacle. In a snowstorm also, a thick layer of snow might give the balance in favour of the arm. To allow for these contingencies the original counterbalancing effort of the spectacle should not be less than equivalent to 100 in. lbs. in excess of the turning moment of the arm. That is equivalent to a weight of 5 lb. at 20 in. from the centre of the spindle.

For these weights the spectacle casting, *without* the glasses in it, should be taken, and the turning moments should be calculated when the arm is in the DANGER position. When it is lowered the counterbalancing effort becomes greater, and, as a rule, is at a maximum at an inclination of 60 degrees, depending on the precise positions of the centres of gravity of the arm and spectacle. It is necessary to fit a stop of some sort on the post so that the arm will not fly be-

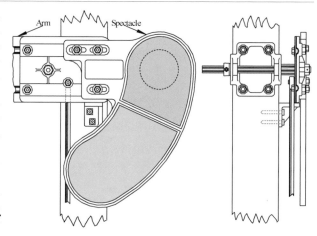

Figure 63
Spindle clear of the post, carried in cast-iron bearing

yond the horizontal position should the connecting-rod break. This stop is either fitted to engage with the wood arm itself, by being placed immediately above it near the bearing, or some portion of the spectacle casting comes in contact with a projection fastened to the post. If the stop engages with the arm direct, the arm soon becomes very badly worn and unsightly at the point of contact. If a metal-to-metal stop is employed no such trouble occurs, but a rather severe blow is liable to be given. Probably the best practice is to provide a wood or other soft material stop to engage with a portion of the spindle casting or spectacle, so arranged that the stop can easily be renewed when worn.

With British Standard arms, which are of the "Upper Quadrant" type (i.e., in which the "Clear" indication is given by raising the arm to an angle of 45 degrees above the horizontal) (see Fig. 60), any additional weight on the arm, due to snow, etc., assists the arm to return to danger. This feature constitutes one of the great advantages of this type of fitting, and, in fact, the Upper Quadrant arm, weighing 45 lb., has a considerably greater torque than any existing Lower Quadrant arm, which weighs from 60 to 90 lbs. Upper Quadrant fittings have also the advantage that they permit the use of posts 3 ft. shorter than the equivalent Lower Quadrant fittings; this is of considerable importance in the case of doll posts on brackets or gantries. British Standard arm fittings are provided with a metal-to-metal stop in which a spring is incorporated to reduce the effect of the blow (see Fig. 60).

The angle to which lower quadrant arms move when in the CLEAR position varies with different companies, but a very fair average is about 60 degrees. It is generally taken that the maximum inclination an arm can have, and still be considered a proper DANGER indication, is about 5 degrees; from this angle until the arm moves to about 35 degrees the signal should be considered as *indistinct,* and from 35 degrees to about 60 degrees a correct CLEAR signal is given (Fig. 65).

It is current practice to have two standard sizes of spectacles only; one the full size for MAIN LINE signals, and the other small to suit small arms of about 3 ft. length. The green glass used in spectacles should be of a decidedly blue shade,

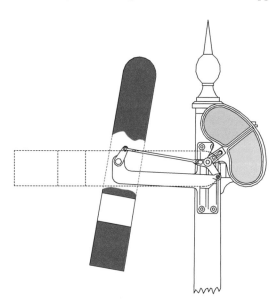

Figure 64
Spindle carried on long bracket

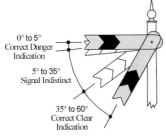

0° to 5°
Correct Danger
Indication

5° to 35°
Signal Indistinct

35° to 60°
Correct Clear
Indication

Figure 65
Angles of working

and as the ordinary light used is yellow, this gives a very good green signal. If the glass is a light green, the yellow light produces a very poor signal light.

The glasses are held in the casting by means of putty, but putty alone is not sufficient, and it is necessary that holes should be drilled round the edge of the spectacle, and either wood plugs driven in or small split pins used.

There are three common arrangements for placing the arm with relation to the post:

1st. The arm is fitted to work inside the post, a slot being cut in the post for that purpose, and the spindle going through the centre of the post (Fig. 66). This involves a costly post and tends to weaken it. It also involves two plate bearings, one on either side of the post, for the spindle to work in. Moreover, the spectacle has to be fitted separately to the spindle, and has a tendency to work loose.

2nd. The arm works outside the post, but the spindle goes through the centre of the post. This involves two plate bearings for the spindle (Fig. 67), but allows the spectacle to be fitted with bolts to the main signal arm casting to which the spindle is keyed. With this arrangement the spectacle must be comparatively heavy to counterbalance the arm, as the spectacle has its centre of gravity close to the spindle, whereas the arm has its centre of gravity nearly half way along the arm (see Fig. 62).

3rd. The arm works outside the post as in the last case, but the spindle is fixed clear of the post (Fig. 68). This has all the advantages of the 2nd arrangement; and it not only allows the bearing to be in one casting, but has the advantage that the centre of gravity of the spectacle is further from the spindle, whilst the centre of gravity of the arm is nearer, thus allowing a lighter casting for the spectacle (see Fig. 63).

With this arrangement the areas on each side of the spindle are about equal, and any snow collecting on the arm would be partially counterbalanced by the snow on the spectacle; so that it would be practically impossible for the snow on the arm to overcome the counterbalancing effort of the spectacle. The only drawback is that the spectacle glasses have to be rather large owing to the distance of the centre of the lamp from the spindle.

This arrangement (of the spindle fixed clear of the post) is adopted in the British Standard design, in the case of which, however, as it works in the upper quadrant, the lamp is placed on the same side of the post as the spindle bearing, thus reducing the required size of the spectacle glasses. This arrangement affords the additional advantage that, as the fittings (lamp, spindle, etc.) are all on the same side of the post, no difficulty is experienced in applying them to different thicknesses of post, which was a distinct disadvantage of some older types in cases where a lower repeating arm had to be fixed (see Fig. 37).

When a spindle is situated approximately at the centre of gravity of the arm, the spectacle must be fitted to a separate spindle bolted to the signal post, with a connecting rod between the arm and the spectacle (see Fig. 64). This arrangement makes it absolutely impossible for any weight of snow on the arm to cause it to droop, but there are more moving parts, and consequently this type costs more to manufacture and maintain, and is slightly harder to work.

In addition to equipping the signal with a front spectacle to show red and green lights, a

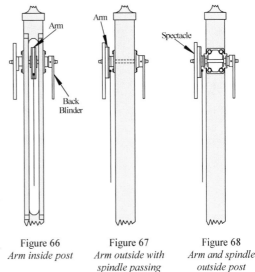

Figure 66
Arm inside post

Figure 67
*Arm outside with
spindle passing
through centre of post*

Figure 68
*Arm and spindle
outside post*

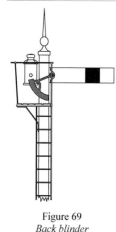

Figure 69
Back blinder

back spectacle (or "Blinder" as it is sometimes called) requires to be fitted (Fig. 69). This is simply a piece of sheet iron (or a casting) fitted to the spindle so that when the arm is at Danger a small white back light can be seen by the signalman, but when the arm is inclined to more than about 5 degrees the light is obscured. The back light arrangement is only used when the face of the signal cannot be seen from the signal box. The reason why the back light is made visible only when the signal is at danger, is because it is essential that the signalman shall be able to satisfy himself that a proper danger signal is being exhibited. Should the light go out when the arm is horizontal the signalman, not seeing the light, knows either that the light is out, or that the arm is inclined more than 5 degrees, and in either case the signal needs instant attention. Should the light go out when the arm is lowered for the "Clear" signal the signalman is not informed of it, but if a driver overlooks the signal owing to the light being out he simply proceeds, which is exactly what he is told to do by the arm being lowered. Should he notice that the light is out, he would take means to inform the signalman, and the worst that could happen would be the stopping of the train when not required, which is on the side of safety.

ELECTRIC REPEATERS

WHEN a signal is out of the signalman's sight, some device is necessary to indicate to him when the signal is at DANGER and when it is at CLEAR. The usual method of repeating a signal arm is by fixing a commutator on the signal post, the commutator being worked by the arm. The contacts of the commutator are so arranged that when the arm is within 5 degrees of the horizontal, an electric current is sent in one direction along the line wire to an indicator in the signal box. When the arm is inclined more than the prescribed amount, and until it reaches about 35 degrees (which latter can be considered a correct CLEAR signal), no current is sent along the line wire, and when the arm is between about 35 degrees and fully off, a current of opposite direction is sent along the line wire (Fig. 70).

At the signal box the indicator consists of a magnetic needle suspended in the field of an electric coil; the needle is connected to a spindle, and a miniature representation of a signal arm is

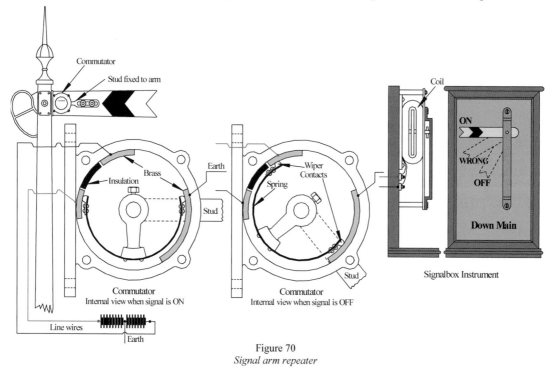

Figure 70
Signal arm repeater

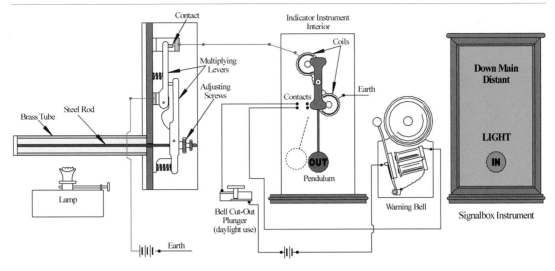

Figure 71
Signal light repeater

also attached to the same spindle.

When the current flows in the first-mentioned direction the magnetic field set up round the needle urges it in a direction which causes the miniature arm to take up a horizontal position (a small stop pin prevents it from moving beyond the horizontal position). When no current flows through the coil, the needle returns to a neutral position by gravity, and the arm in this position indicates OUT OF ORDER, or WRONG. When the current flows in the reverse direction, the needle is urged in the opposite direction, which places the miniature arm fully off, indicating CLEAR.

This instrument is termed a "three-position" instrument, as it indicates the signal when "ON," "OUT OF ORDER," and "OFF." It will be noticed that any failure of current indicates "OUT OF ORDER."

Some repeaters were employed at one time which indicated "ON" when no current flowed, and a current was sent only when the signal was at "clear," this being termed a "two-position" instrument.

Repeaters are most commonly fitted to distant signals, as these signals are often placed over 1,000 yards from the signal box, and this great distance in many cases places them out of the signalman's view; when the distant signal is the sole arm on the post only one repeater is required, working in conjunction with the arm.

It is the practice on many railways, in cases where the distant signal is fitted lower down on the post carrying the starting signal for the box in the rear, to repeat the slot lever in addition to repeating the arm of the signal. The reason for repeating the slot lever is to inform the signalman how his portion of the signal is working, as should he pull his lever, and the arm repeater show that the arm did not move, he would be uncertain as to whether he had performed his portion of the work correctly, or whether the signalman in the rear box had his starting signal in the danger position thus holding the distant arm at danger. Also it is most important that the signalman shall know that the arm in returning to danger has acted in obedience to his lever. If his signal wire should become "jammed" when restoring his signal lever to the normal position, the slot lever would remain in the clear position, but the arm would be at danger as soon as the starting signal assumed that position; then when the starting signal is next pulled to clear, the distant signal would follow to clear, although the signal lever is in the normal danger position.

It is also necessary to repeat the light of a signal in addition to repeating the arm, when it cannot be seen by the signalman operating it, or an adjacent signalman. A light repeater works on the principle that a lamp when alight gives off heat, and that different metals (such as iron and copper) have different coefficients of expansion. Two dissimilar metals are coupled together in such a manner that any variation of temperature causes one piece to project beyond the other; the difference is very small, but by a system of levers this small difference is amplified sufficiently to make or break an electric contact. Fig. 71 shows one form of light repeater. The tube is made of brass, and the inside rod of steel. Both are fixed at *one* end, and the free end of the steel is connected to a lever; this lever rests on another one, to the end of which a contact is fitted. The device is adjusted so that when *cold* the contact is *broken,* but when the lamp is alight the heat causes the brass tube to expand more rapidly than the iron rod, thus giving a relative motion between them, and so moving the levers sufficiently to make up the contact; this sends a current along the line wire to the signal box. Should the lamp go out, the tube and rod return to the normal temperature and the contact is broken.

At the signal box an instrument is fixed, which, when a current is flowing, indicates "LIGHT IN," and any cessation of current, from whatever cause, allows the indicator to return by gravity to its normal position, which indicates "LIGHT OUT." When the indicator falls to the "LIGHT OUT" position, it makes a contact, which closes an independent bell circuit, and the bell rings to call the signalman's attention to the light being out. To prevent the bell from ringing during the day time, a small switch is provided, which completes the bell circuit when required.

Where the signals are electrically lighted, a small glow lamp—fixed in the signal box—is included in the lighting circuit of the signal to be repeated. Any failure of light in the signal is indicated by the glow-lamp going out.

When the distant signal is combined on the same post as the starting signal of the rear box, the signalman at that box is responsible for observing the lights of both signals, and the distant signal light is not then repeated to the box working the signal unless the rear box is switched out during hours of darkness.

COUNTERBALANCE LEVERS

THE connection between the signal box and the signal itself is made of wire, and arranged so that pulling the wire inclines the arm to the CLEAR position. The usual method adopted is to attach the wire to one end of a lever fitted to the signal post (Fig. 72). From the lever a rod is taken to a pin or stud on the signal arm, so arranged that when the wire is pulled the rod moves up, and being connected to the right-hand side of the signal arm spindle, the arm is thereby lowered. When the wire is loosened again the arm should go to danger and the wire is at the same time dragged back. The weight of the down rod and lever alone is not always sufficient to drag back the wire, so it is usual to place a weight on the tail of the lever to assist in dragging back the wire. It should be observed that the true function of the weight is to *pull the wire back* to its normal position, and NOT to put the signal to danger. Lower quadrant spectacles should always be sufficiently heavy to put the arm to danger even when loaded with snow, but unless the counterbalance weight is sufficiently heavy to pull back the wire, the arm will, of course, remain in the CLEAR position. When the wires are loaded with snow, and every pulley wheel between the signal box and the signal partially choked with snow also, the weight on the counterbalance lever is not always sufficient to overcome the additional friction introduced, and then, of course, the signal never properly goes to danger. The further the signal is from the signal box the greater is the effect of the snow on the wires and round the pulley wheels. When a very severe storm is encountered the effect is too great for any reasonable amount of weight to overcome. Under these conditions the only practical thing to do is to arrange for the signal being tied, or otherwise secured, in the danger position.

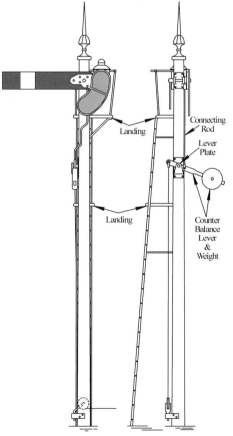

Figure 72
Signal with counterbalance lever and weight

With the new British Standard (Upper Quadrant) Arms, however, a metal-to-metal stop is incorporated in the spectacle casting itself, and so it becomes possible to utilise a wire for connecting the arm to the lever plate. In the case of signals not far from the signal cabin the weight of the steel arm is often sufficient to pull the wire back and a lever plate may not be required. With this design of fitting, also, if the wires are loaded with snow it is probable that a quantity will have adhered to the arms, thus increasing the weight available for pulling back the wires, so that the effect of snow on this type of arm is, to some extent, self-compensating.

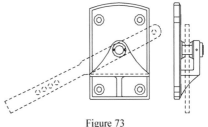

Figure 73
Lever plate with integral stops

The usual sizes for the counterbalance levers, etc., are as follows :

Length from centre of lever bearing to centre of pin connected to wire, $5\frac{1}{2}$ in. ; this gives about $4\frac{1}{2}$ in. travel for the signal wire; from centre of bearing to extreme end of lever, 20 in. The weight of the counterbalance casting varies from 20 lb. for signals about 300 yards from the signal box, up to 40 lb. for signals about 1,000 yards from the signal box, but this depends very much on the run of the wire. Several holes are drilled in the lever to allow of the distance of the counterbalance weight from the centre being adjusted after the signal has been erected and the wire connected up. The size of the connecting rod (commonly called the *down rod)* varies from $\frac{5}{8}$ in. for short rods to $\frac{3}{4}$ in. for long rods. If the rods are long it is necessary to fit guides on the post to prevent their bending.

With lower quadrant arms the position of the counterbalance lever on the post varies on different railways. They were in many cases fixed close to the bottom of the post, within easy reach of the signal linemen for examination and cleaning purposes. This, of course, necessitates a long down rod, and should the post be a very high one the down rod becomes very heavy. It is, however, sometimes possible to connect the wire direct on to the lever, and so save a pulley wheel at the foot of the post.

If the lever is placed about 4 ft. 6 in. below the arm, it is possible to adopt a standard length of down rod, and there is no friction incidental to the use of rod guides. This arrangement on the whole gives a very good working signal when first erected, but unfortunately, when the counterbalance levers are high up the post, they are not easily accessible for inspection, and the lineman has to climb the post every time the bearings require oiling. If the bearings cannot be easily reached from a ladder a small landing should be fitted at a convenient height to enable oiling and cleaning to be carried out in safety (see Fig. 72).

With British Standard (Upper Quadrant) arms, the rod or wire moves *downwards* to incline the signal arm so that a long and weighty down rod would outbalance the weight of the arm and so must not be permitted. The standard down rod is not more than 11 ft. 6 in. in length.

There are many different types of bearing plates for the levers, and if only one lever has to be accommodated there is no special difficulty in obtaining a good design. The plate must be fitted

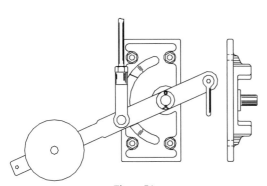

Figure 74
Lever plate with stops projecting from face

with two stops, in addition to the bearing, to prevent the counterbalance lever moving too far in either direction. Possibly the most convenient pattern to adopt is one with the pin for the bearing in double shear (Fig. 73), and not overhung. In this pattern the stops are formed by the lever striking the body of the casting. With the pin overhung the stops take the form of projections from the face of the plate (Fig. 74), and unless these are very strong there is a liability of their breaking off.

Where there are two arms on one post, such as in the case of a starting and shunt signal, with both signals reading in the same direction, the lever plate, as a rule, is made so to receive the counterbalance levers for both arms with a washer between them. In the case of home and distant signals on the same post, it is necessary to control the distant arm so that it cannot be cleared until the home arm has been cleared; this arrangement being termed "SLOTTING THE DISTANT SIGNAL."

SLOTTING

THE term *slot* is a survival from the time when the signal arms were not counterbalanced by their spectacles; an early device for controlling the distant arm being as follows (Fig. 75):

The distant signal down rod, instead of having a common single joint to fit on to the distant arm driving stud, has a slotted joint so arranged that when the counterbalance lever is normal

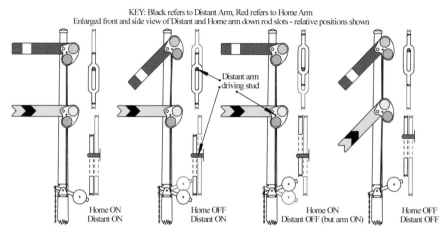

KEY: Black refers to Distant Arm, Red refers to Home Arm
Enlarged front and side view of Distant and Home arm down rod slots - relative positions shown

Distant arm driving stud

Home ON
Distant ON

Home OFF
Distant ON

Home ON
Distant OFF (but arm ON)

Home OFF
Distant OFF

Figure 75
Early method of signal slotting

the arm is at danger; but when the counterbalance lever is raised the slot moves up, and if there is nothing in addition to this slotted joint to prevent the arm from lowering, it will move down as the down rod moves up; but if the arm is held by any other means in the danger position, the slot simply moves up without affecting the arm. The down rod of the home arm also has a slot formed in it to fit on to the same driving stud as the distant signal down rod, in such a way that unless the home arm is lowered the slot in the down rod of that signal prevents the distant arm from falling to the OFF or clear position, although the distant signal counterbalance lever be raised. It will be seen that with this arrangement of slotting, both counterbalance levers have to be up before the arm can be lowered, and either of the counterbalance levers going back to the normal position carries the distant arm with it.

When the Board of Trade made the ruling requiring the spectacles to counterbalance the arms, this method of slotting was no longer admissible, as it relied on the *weight of the arm* to put the signal to the OFF position, and for the *weight of the counterbalance* weight and lever to put it to *danger*.

The device generally in use is a modification of the above system. The spectacles are made of the standard pattern and effectively counterbalance the arm. There are no slots in the down rods, but an additional so-called "SLOT" lever is placed between the ordinary counterbalance levers with the long end pointing away from the other levers. The down rod of the distant arm is connected to this lever, so that when the long end moves *down* the arm follows and takes up the CLEAR position (Fig. 76). A small counterbalance weight is fitted to this lever, sufficient to overcome the weight of the distant arm and spectacle. A stud or cross-piece is fastened or welded to the slot lever, and lies across the counterbalance levers, in such a way that unless both these levers are raised it is impossible for the slot lever to fall and pull the distant signal arm with it. If the cross-piece (Fig. 77) is on the same side of the bearing as the down rod it is placed *below* the levers, so that it can only rise when both are up. If, however, it is on the opposite side of the bearing, it is placed *above* the levers, so that it bears down on them, and cannot fall unless both ends are down.

Should the distant signal require to be controlled by more than the top arm, all that is necessary is to place another ordinary counterbalance lever on the same bearing as the others, and arrange that the cross-piece of the SLOT LEVER bears across all the counterbalance levers.

There must be as many counterbalance levers as there are controls required, with the slot lever cross-piece bearing on each one. With this method there is, theoretically, no limit to the number of controls which can be effected. Practically the limit is determined by the size of the lever plate and the feasibility of making a strong cross-piece to lie across *all the counterbalance levers.*

This type of slot is the most convenient to fit up and the easiest to work, and if the type of lever plate is such that the bearing pin is in

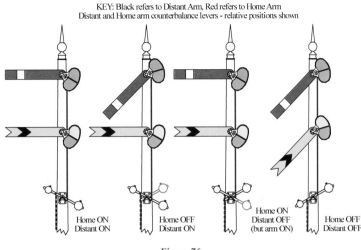

Home ON
Distant ON

Home OFF
Distant ON

Home ON
Distant OFF
(but arm ON)

Home OFF
Distant OFF

Figure 76
Common slot signal

double shear, and the stops for the levers a part of the main casting and not mere projections therefrom, four separate controlling levers (five levers including the slot lever) can be employed without difficulty. If, however, the bearing pin is in single shear, and the stops for the counterbalance levers are projections from the main casting with more than three controlling levers (four including the slot lever), the pin bearing becomes rather long, and the stops are liable to fracture in service.

There is one theoretical objection to this type of slotting, *viz.*, should the cross-piece on the slot lever *break*, the lever would fall and carry the distant signal arm with it, so that in this case a *failure of connections* would cause the signal to go to CLEAR, which is *not* in accordance with the Ministry of Transport Requirements. In practice, therefore, great care is taken in designing the slot lever and cross-piece so that it will be so strong that whatever portion may break the slot lever cross-piece will NOT fail. This type of slot is commonly called a "DROP-OFF" SLOT owing to

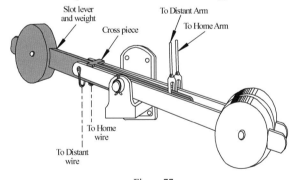

Figure 77
Common or "drop-off" slot

the fact that the weight on the slot lever in falling or dropping pulls the distant arm to the CLEAR, or off position.

There are several types of "PUSH-OFF" SLOTS which are sometimes used. They have not the theoretical objection of the DROP-OFF slot, but they have the practical objection that they do not work so easily, and it is not easy to arrange for more than two controlling levers. Should more than two be required they have to be arranged one above the other in series.

The general principle of most "PUSH-OFF" slots is as follows:

Taking a two-lever control as before, three levers are required, a slot lever being placed between the two control levers, with the weight (if any) assisting to put the arm to danger. In addition to the cross-piece fixed to the slot lever, a movable crank (see Fig. 78) is fitted, with the ends of the crank resting on the control levers situated on either side. When one control lever is pulled, it simply pushes over the small crank on to the other control lever. When that control lever is in turn pulled, it tries to force the crank back on to the first lever, but as this lever is already pulled up, the effect is for the crank to rise with the second lever and carry the slot lever with it. Whichever lever is put back first it allows the crank, and consequently the slot lever, to go back to the normal position along with it. In some varieties of this slot, instead of the small moving crank being fitted directly on to the slot lever, it is fitted to a sliding rod (see Fig. 79), which in turn is fitted to the slot lever; then, instead of the crank resting directly on the control levers, it rests on studs riveted to two slides which are driven by the control levers. The action is precisely the same, as whichever lever is pulled up first drives its sliding piece and, by means of the stud riveted thereto, forces the crank on to the other stud. When the next lever is pulled up, it drives its sliding piece up, and its stud, not being able to force the crank over against the first stud, raises the crank which pulls up the sliding piece attached to the slot lever, so lowering the arm. Fig. 80 shows another variety of "push-off" slot. It's method of operation is probably self-evident from the illustration.

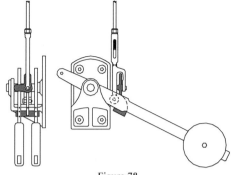

Figure 78
"Push-off" slot with direct acting crank

The objection to any of these arrangements is that the friction of the moving parts is comparatively great. The lever first pulled does little work on the slot, as it simply moves the crank or roller over, but the second lever pulled has the whole slotting device to operate, and the signal arm to move in addition. If the first lever to be pulled is the one working the distant lever, little difficulty is experienced, as generally the starting signal is sufficiently near the box to enable the signalman to pull off both it and the slotting without much trouble.

If, however, the starting signal lever is first pulled, when the signalman comes to pull the distant lever (especially if the signal is at its usual 1,000 yards or so from the box), it is almost impossible for him to operate the control lever. Where more than two control levers are required, it is quite impossible for a signalman 1,000 yards away to pull the signal off should he be the last man to pull, owing to the amount of friction involved in the arrangement.

With either of these types of slotting, should only one arm be on the signal post, and that arm required to be controlled from two

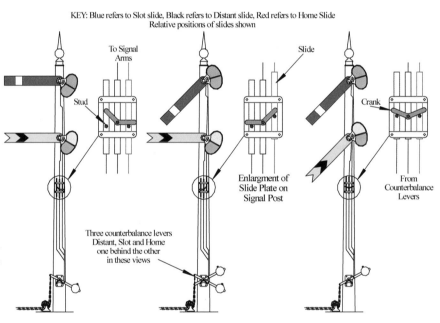

KEY: Blue refers to Slot slide, Black refers to Distant slide, Red refers to Home Slide
Relative positions of slides shown

To Signal Arms
Slide
Stud
Crank
Enlargment of Slide Plate on Signal Post
From Counterbalance Levers
Three counterbalance levers Distant, Slot and Home one behind the other in these views

Figure 79
"Push-off" slot with crank and slides

signal boxes, the same apparatus would be fitted as for a home and distant signal on one post, with the difference that the down rod from the one counterbalance lever would be omitted; thus none of the counterbalance levers would be connected directly to the signal arm, the down rod from the signal arm being connected to the slot lever.

Another method of effecting a *slot* is by means of pulleys and the signal wires. Instead of the signal wire being taken direct from the signal box to the counterbalance lever on the signal post, the wire from one box is led round a pulley wheel (which pulley wheel is allowed to slide) back to the other box (Fig. 81). From the sliding pulley wheel a wire is taken to the counterbalance lever

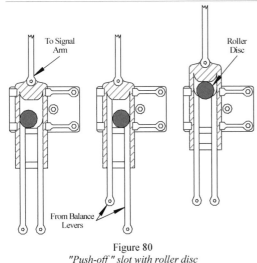

Figure 80
"Push-off" slot with roller disc

of the signal requiring to be controlled. The method of working is as follows: the wires are left slack, so that when the first man pulls his end of the wire he simply tightens the wire, but not sufficiently to pull off the signal. When the second pulls the motion is sufficient to operate the signal. It will be seen that it takes two men to pull off the signal, the first man simply taking up the slack wire and the second man actually working the signal; then, whichever man lets out the wire first, it causes the signal to go back to the danger position. This device would be a very convenient method of carrying out slotting if it were not for the difficulty experienced in regulating the length of the wires owing to expansion and contraction due to varying temperatures during the course of a day and night. Unless the length of the wires is carefully regulated, it would be possible with an abnormally tight wire for the first man pulling to cause the arm to move out of the correct danger position.

A method of signal control which in some ways resembles "slotting" is called "reversing" or "disengaging." These terms cover both electrical and mechanical devices whose function is, broadly speaking, to disconnect the signal from the lever or wire which operates it, so preventing it being cleared if at danger, and allowing it to return to danger if in the pulled position.

Reversers or Disengagers may be arranged to be operated by the passage of a train, or by the pulling of a lever in a locking frame, in which latter case they approximate in effect to the ordinary "slot." They have this difference, however, that with a slot the operation of clearing the signal is carried out by the *last* pull, irrespective of the order in which the controls are operated; with a "reverser" or "disengager" control, on the other hand, in order to pull off the signal arm the engaging lever must be operated *first* and the operating lever *last;* if these operations take place in the reverse order the signal will

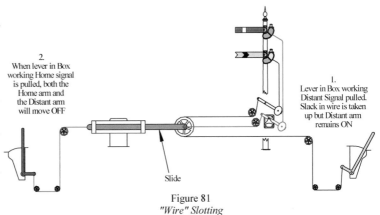

Figure 81
"Wire" Slotting

not be cleared. Thus, if one signalman "slots" a signal worked by another, either of them can put the signal to danger and pull it off again; whereas if one signalman "disengages" a signal worked by another he can put the signal to danger (by disengaging it) but cannot pull it off again. This distinction becomes important in the case of inter-controlling by signal boxes which close (or "switch out" as it is termed) at different times.

REVERSERS

IT is sometimes desirable to arrange that the train passing a certain point shall automatically place the signal behind it to danger. This can be done mechanically, as shown in Fig. 82. The signal wire is taken to a cross-bar instead of directly on to the counterbalance lever, and the opposite end of the cross-bar is connected to a catch, this catch being operated by a treadle placed by the side of the running rail, so that a train passing over the treadle releases the catch and allows the signal arm to go to danger. Either the flanges of the wheels operate the treadle, or, preferably, the depression of the rail is used for its operation. To pull the signal off again, the lever in the signal

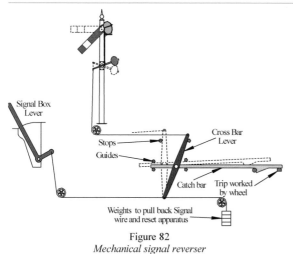

Figure 82
Mechanical signal reverser

box has to be replaced and pulled a second time. It is, however, usual to effect the reversing of a signal by some electrical method, the device generally adopted being "Sykes' Reverser" (commonly called "Sykes' Banjo").

This is fixed on the signal post, as shown in Fig. 83, and acts in a similar manner to the mechanical reverser just described.

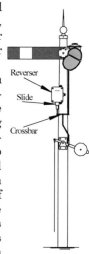

Figure 83
Arrangement of Sykes Reverser on post

The signal wire is taken directly to the counterbalance lever, but instead of the signal connecting rod being fastened to the counterbalance lever, it is connected to the cross-bar. The opposite end of the cross-bar is connected to the slide of the reverser, the centre connection being taken to the counterbalance lever. When the signal is pulled off, the slide of the reverser endeavours to move UP; it is, however, prevented from rising by the roller on which it presses (see Fig. 84). This roller is fastened to an arm which is prevented from swinging by a small catch. To reverse the signal the hammer is allowed to fall on the rod connected to this catch. When this occurs the catch is forced clear of the arm, the slide is then able to rise, pressing on the roller, and so causing the arm to swing to the right. As the arm swings the cam-shaped portion at the top of the arm engages with the hammer and restores it to its normal position. When the slide of the reverser moves up, it allows the connecting rod of the signal to move in the opposite direction, thus putting the signal to danger. During this motion the cross-bar swings with the pin of the centre connection for its bearing.

The hammer is held up by a lever connected to the end of the armature of an electro-magnet.

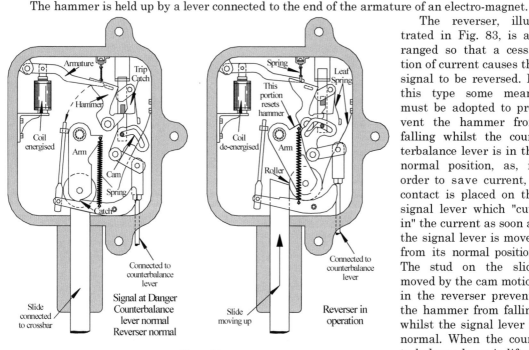

Figure 84
Sykes Reverser arranged so that a cessation of current reverses the signal

The reverser, illustrated in Fig. 83, is arranged so that a cessation of current causes the signal to be reversed. In this type some means must be adopted to prevent the hammer from falling whilst the counterbalance lever is in the normal position, as, in order to save current, a contact is placed on the signal lever which "cuts in" the current as soon as the signal lever is moved from its normal position. The stud on the slide moved by the cam motion in the reverser prevents the hammer from falling whilst the signal lever is normal. When the counterbalance lever is lifted, the hammer moves forward slightly, until it is stopped by the armature lever (the lever is in the correct position for stopping the hammer only when the coil is energised). The trip catch piece moves forward at the same time as the hammer, and is caught by the back of the hammer; when the hammer falls it moves slightly forward again. While the hammer is being re-set, the catch is moved, and when the hammer is completely re-set the catch flies forward and holds the hammer *up* until the counterbalance lever resumes its normal position. The stud now holds the hammer up again, and the catch piece is then returned to its normal position.

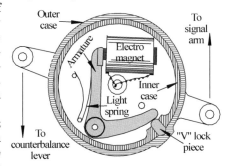

Figure 85
Semi-rotary reverser for signals

Very often it is arranged that a momentary current causes the hammer to fall. The trip catch portion is then dispensed with and the armature is placed below the electro-magnets; the lever holds the hammer in position until a momentary current lifts it and allows the hammer to fall.

Reversers of different patterns and designed on different principles are in use; one of the simplest forms is shown in Fig. 85. An electro-magnet together with an armature is fitted inside a rotatable case. The case consists of two portions, each capable of rotating partially on the same centre; one portion is connected to the signal arm and the other is connected to the counterbalance lever. When an electric current flows through the coils the two portions act together, and if the counterbalance lever is raised the signal arm is lowered; when the current ceases to flow through the coils the armature is forced back and the two portions are free to move independently. When this occurs there is nothing to hold the arm in the "off" position, it therefore goes to danger owing to the counterbalancing effort of the spectacle and the weight of the connecting rod.

By reason of the complicated design of most reversers, with the consequent necessity for specially careful maintenance if "wrong-side" failures are to be avoided, there has lately been a noticeable tendency for "all-electric" operation (by primary batteries, or otherwise) to be installed in the case of signals on which reversers would formerly have been fitted.

TREADLES

CONTACTS operated by a train for making or breaking electric circuits are generally termed treadles.

Practically all treadles depend on the fact that when a loaded wheel passes over a rail the rail is depressed. This depression is very much intensified by some means, and the motion obtained is employed to "make" or "break" an electric circuit. The make or break device may be a simple "wipe" or "rubbing" contact, as used in a single repeater (see Fig. 86), or it may be a mercury contact, also shown in Fig. 86. With the mercury contact, mercury is poured into an iron chamber and the insulated wire fitted inside the chamber so that when the device is in the normal position the wire is clear of the mercury, thus breaking an electric circuit, and when it is in the reverse position the wire dips into the mercury, completing the circuit.

The iron chamber referred to is fitted with bearings, one end of it engaging with a multiplying lever which is operated by the deflection of the rail.

Another device using mercury is Siemens' treadle (see Fig. 87).

In this treadle a comparatively large body of mercury is placed in a container below the rail; the top of the container is of corrugated plate, which is allowed to press against the underside of the rail by means of a plunger. From the container a small bore tube is taken which ends in a cup at a higher level than the container. When a train passes over the treadle the mercury is forced from the container up the tube into the cup, and so comes in contact with a wire, making up a circuit. The mercury is allowed to return by means of a duct to the main container after the train has passed, the mercury acting like water in a hydraulic ram.

It can also be arranged that the wire dips into the mercury normally, and that the passing of the train forces the mercury up the tube into the cup; when the wheel is off the treadle the mercury recedes down the tube, thus breaking the circuit momentarily.

The treadles of whatever type are in most cases attached entirely to the rail, the weight of the train causes the rail to deflect between the treadle supports, and this deflection alone must be

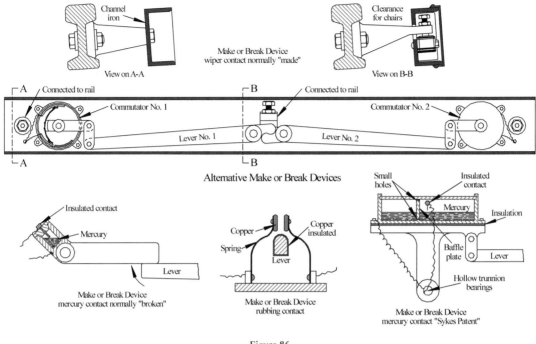

Figure 86
Treadle contatcs

sufficient to work the treadle. It is important to see that the deflection is sufficient to work the contacts, but at the same time it should not be possible to operate the treadles with a light plate-layer's bogie, or anything but a heavy vehicle.

These devices are used for high speed running where momentary makes or breaks only are required. Where it is required to indicate the presence of a standing train, and maintain a signal at danger, a balanced bar is fitted. Fig. 88 shows an end view of a balanced bar. The flange of the wheel presses the L iron down, and so makes or breaks an electric circuit. When the wheels are off the bar it returns to its normal position by means of gravity, the I section iron counterbalancing the L iron portion of the bar. The contacts worked by this bar are, as a rule, of the simple "slide" or "wipe" pattern. Balanced bars of this type are commonly used at fouling points to reverse, or keep at danger, a conflicting signal, and they are also connected up to electric locks preventing certain signal, or point, levers from being pulled whilst a train is on the bar.

In some systems of signalling the balanced bar is used instead of a facing point lock-bar, the facing point bolt being worked along with the points by means of an escape crank. When a wheel depresses the bar, an electric lock holds the point and bolt lever from being moved, and when there are no wheels depressing the bar the points may be moved as required.

In order to minimise the shock of the wheel depressing the bar, the extreme ends of the bar are bent down to form a ramp; this causes the bar to be depressed more gradually.

Fouling bars are chiefly used at junctions where the fouling point cannot be easily seen from the signal box. There are also cases where the fouling point of a siding cannot clearly be seen from the signal box, and a fouling bar, placed along the switch blade or

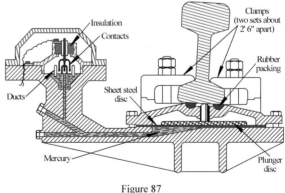

Figure 87
Siemens' Treadle

stock rail (outside), is made to lock the point lever, thus preventing the catch points being moved until the train is completely inside the siding. Bars of this description are also used where the view of vehicles standing in a platform, or dock, is bad. The bars in this case operate an indicator in the signal box, and at the same time lock up all incoming signals, except calling-on signals, so far as that platform or dock is concerned. Where fouling bars are used for this purpose they should be spaced along the platform line at such a pitch that

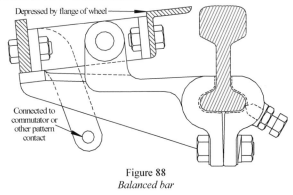

Figure 88
Balanced bar

no vehicles can be left standing between the bars. On long platform lines this would necessitate a large number of bars, and to economise, the bars are often placed at greater intervals, in which case there should be instructions issued that all vehicles left in docks so equipped must be left standing on one or more bars, and some form of indicator should be fixed to mark the position of the bars, so that the trainmen can see, without difficulty, where to leave the vehicles, and be certain that a bar is depressed. Fouling bars, like lock-bars, must be longer than the greatest wheel base of the vehicles using the line.

TRACK CIRCUITS

OF late years the cost of providing and maintaining treadle bars has led to their replacement in many cases by Track Circuits. Track Circuits are commonly employed for the purpose of indicating the presence of a train on a portion of line. To equip a portion of line for this purpose it is first necessary to insulate the portion of line being treated from the remainder of the line. Insulated joints, consisting of fibre placed between the fishplates and the rails, fibre ferrules round the fishplate bolts, and a piece of fibre between the two adjacent rail ends, must be inserted on each line of rails at both ends of the section. Bond wires must be fastened from rail to rail across the fishplates in order to increase the conductivity of the rail joint. At one end of the section so insulated and bonded, a battery is connected, and at the opposite end of the section a relay is connected. The current from the battery flows along one of the bonded rails to the relay and back along the opposite rail to the battery; this completes a track circuit. When no train is in the track-circuited section the relay is "energised," and the armature of the relay is held up against the electromagnet. When a train enters the section, if the wheels are metallic throughout, the current from the battery flows through the wheels and axles of the train to the opposite rail, and as the resistance of the wheels and axles of a train is much less than the resistance of the relay, most of the current flows through the former and a very small amount flows through the relay. If the resistance of the relay has been designed to suit the conditions of service, the relay will become "de-energised" when the train is in the section; that is to say, the armature is no longer held up against the electro-magnet. The armature can be made to operate any secondary circuits desired, such as operating a VEHICLES-ON-LINE indicator, lock and reverse signals, etc., etc. Track circuiting can be used instead of treadles or fouling bars; it can also be used to lock a facing point lever when a train is standing foul of the points, and in this case it performs the function of a lock-bar.

In practice the following arrangements are commonly adopted:

- *Bonding.*—There are several varieties of bonds used; some bonds are galvanised iron wires pinned into holes bored in the rails by means of channelled plugs; other bonds are made of stranded wires secured to hollow plugs, which are tightened in the holes in the rails by driving a steel pin through the hollow plug, thereby swelling it into the hole.

- *Batteries.*—Copper sulphate gravity pattern cells were very commonly used for track circuits as their voltage and internal resistance make them very suitable for this work. The E.M.F. is approximately 1 volt, and with two cells in parallel the internal resistance is about 1 ohm. These cells are about 6 in. diameter and 8 in. deep, and will work for one month under ordinary conditions without renewal. The elements are copper and zinc.

Recently caustic soda cells, with copper oxide in place of the zinc element, are much used. The E.M.F. is very low —about 0.66 volt—but the internal resistance is also extremely low, so that with a short circuit a large current flows from the battery. In practice an external resistance of some description is added for the purpose of preventing an undue amount of current being taken from the battery when a train is on the track circuit; further, the external resistance improves the shunting of the relay. These cells remain in service for long periods without renewal. Storage batteries have also been used for track circuit work, and, like the caustic soda batteries, these require an external resistance added to the circuit.

As the insulation of the rails from earth is comparatively poor and varies from day to day with the weather, low voltages only are suitable; with high voltages the leakage would be too great. Even with the low voltages used only a fraction of the current output of the cells reaches the relay if the track conditions are bad. If the track is naturally dry and well ballasted and drained, it is possible to have track circuits upwards of 1,000 yards in length, but with poorly drained track and with low insulation from earth 1,000 ft. may not be a workable length; as if the amount of current leaking from rail to rail via the sleepers and ballast is too great, sufficient current cannot flow through the relay to energise it.

Alternating current supplied from power mains is used when electric traction instead of steam is employed, and is preferable to direct battery currents if stray currents are experienced; it is more commonly employed for the operation of automatic power signalling.

- *Track* Belays.—Resistance about 9 ohms, but must be such as to suit the batteries used and the insulation of the rails or "ballast resistance." Contacts on armature for operating secondary circuits are generally carbon to silver. It should be designed to attract the armature when about 0.04 amp. flows through coils, and allow the armature to move away from the coils when 0.027 amp. (or less) flows through coils.

The batteries are placed in "chutes" let into the ground below the frost line in countries subject to severe frosts, or in weather-tight cupboards fixed by the side of the line where severe frosts are not experienced. The relays are generally housed in wooden shelters which must be weatherproof. If the rolling stock is fitted with wood or composition wheel centres, it is necessary to connect the tyres of the wheels with the axles by means of copper wire to ensure good conductivity. Even with properly bonded wheels one vehicle by itself cannot be relied upon to short-circuit the relay effectively. The weight of an engine is, with rare exceptions, sufficient to ensure a good electrical contact between the rails and wheels.

SIGNAL LAMPS

THE majority of signals are lighted by oil at present, although near large towns either gas or electric lighting is often used.

It is impossible to give a general statement as to whether gas or electric light is the cheaper to install and maintain; everything depends on the rates charged. If the rates for simple lighting by the two systems came out equal, electricity would be much cheaper in the long run, as gas very soon causes the lamp cases to deteriorate. With electric light it is possible to switch in or out just when and as required, and it is thus possible to have fewer hours burning with this system than with any other. The lamps also need less cleaning and general attention, so that the labour costs are very much less.

For general use, however, oil lighting is the most economical, and if lamps of the long burning type are employed at busy places the labour costs for attention can be kept very-low. Generally speaking, at ordinary country or way-side stations common oil lamps, attended by the station staff, are the most serviceable.

Whatever the system of lighting employed, there must be an outside lamp case attached to the signal post. This lamp case encloses the light and in most designs contains a lens to concentrate the rays.

The lamp case must be made strong enough to withstand all weathers, and if a flame light is used it must provide adequate ventilation for supplying air and preventing sweating; at the same time it must be so constructed as to prevent a strong gust of wind from extinguishing the light (Fig. 89).

The lamp cases generally are made of either sheet copper, sheet iron, or cast iron. The sheet-iron cases wear out more quickly than those made of copper, but they are cheaper for first cost. The cast-iron cases (providing they are not damaged by ill-usage) last indefinitely. A lid is provided to allow the interior to be put in, and the cases must be fitted with some attachment for fixing to the signal post. Some lamp cases for electric light are made with the lid to open at the ends, and the inside of the lamp case is only large enough to accommodate the electric bulbs at the proper distance from the lens. Usually, however, the lamp cases are made so that in the event of emergency oil interiors can be fitted and the signals lighted by oil. This also has the advantage of keeping

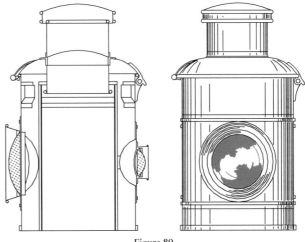

Figure 89
Sheet metal outer lamp case

a standard lamp for all systems of lighting on the line, as the lamps for gas lighting have the burners arranged to lie flat so that an oil interior could be put in should the gas be cut off in case of accident.

The interior case for oil burning contains an oil reservoir and burner, with a means for adjusting the height of the wick when the interior case is closed. Two sides of the case at least must be of glass, as a lamp is fitted with front and back lights. The shape of the interior case should not be square but rectangular (unless some other method is adopted) to prevent the possibility of its being inserted in the outside case with its metal side facing the lenses (Fig. 90). The inner case must be strong enough to allow of its being carried from the lamp room at a station to the signals, which are often about three-quarters of a mile away, in the case of distant signals.

Ample ventilation must be provided, and a lid is fitted on top of a short chimney to prevent the light being blown out when being carried from the lamp room to the signal. It is usually arranged that it is impossible for the lid of the exterior case to be shut down until the lid of the interior case has been opened up, thus ensuring good air circulation. This is effected by making the carrying handle of the inner case to lift the lid when it is pressed down, and unless the handle is pressed down it prevents the lid of the outer case from being closed.

The interior cases are made of No. 24 Standard Wire Gauge, charcoal tin plates, or copper sheets of about the same gauge. Common paraffin oil is used, and the cistern or reservoir usually holds about $1\frac{1}{4}$ pints. The size of the wick is usually $\frac{3}{8}$ in., and burns slightly less than one-third of an ounce per hour (16 oz. = 1 pint) with a "Barton" vapour burner. The flame gives from 1 to 2 candle power.

In common lamps a 6-in. plano-convex lens with $3\frac{7}{8}$ in. radius of curvature is generally employed, the correct distance for the flame from the plane of the lens being about 6 in. If the signal is very high the flame of the lamp is raised above the centre of the

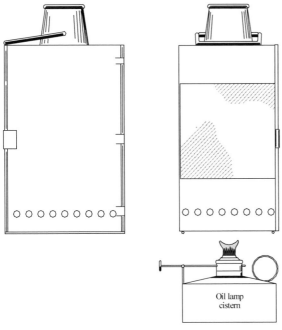

Oil lamp
cistern

Figure 90
Oil burning inner lamp case

lens in order to throw the light down. In the case of a ground signal the centre of the flame is placed below the centre of the lens to throw the rays upwards.

This size of lens suits main line signals, but where short arms are employed it is usual to keep the body and interior of the lamp the standard size and fit a smaller lens about 3 in. or 4 in. diameter. If the standard size main line lamp is very large, it would look very much out of proportion if placed on a post carrying a short arm, in which case a smaller lamp should be adopted.

Of recent years several firms have placed on the market what are known as "Long Burning Lamps." These lamps are intended to burn for about seven days continuously without attention. A good quality oil is required, costing more than common paraffin, but the burners are designed to consume less oil than the ordinary $^3/_8$ in. wick, although the long burning lamp consumes oil continuously, and the ordinary lamp only burns about fifteen hours per day on an average all the year round. The long burning lamps, when of good design, have been found to be less costly in upkeep so far as the oil bill is concerned; and where a large number of signals can be grouped in a convenient area a man and a boy can be set apart to trim and keep them burning, visiting each lamp once per week, at a less labour cost than where each lamp has to be attended to each day. In addition to this, should a sudden darkness come on the lamps are always burning, and there is no necessity for sending out men to light the signal lamps.

The chief features of the long burning lamp are:

1st. A burner which will give out an efficient light for seven days without the wick requiring attention.

2nd. A burner which will give out a good light on a very small consumption of oil.

3rd. A lens or reflector which will grasp the maximum amount of light rays and send them through the spectacle glass.

So far as the third item is concerned, it is obvious that with a common plano-convex lens, the nearer the flat side is to the flame the greater will the number of rays be that are grasped by the lens. But the nearer the lens is to the flame the shorter is the radius of curvature to suit the correct focal length. In practice it is not advisable to use a lens of less focal length than 8 in. where the diameter of the flat is 6 in., otherwise the thickness of the lens greatly retards the light. This can be partly obviated by using stepped lenses, but these again are more difficult to keep clean.

Items 2 and 1 are more difficult of solution, and existing burners are covered by the inventors' patents.

In all signal lamps a small lens for a back light is fitted, the lens being about $^3/_4$ in. to $1^1/_2$ in. diameter. Should the face side of the signal be visible from the signal box the lens is generally covered up with a piece of metal. Where reflectors are used a small hole must be left in them to allow of this lens being illuminated.

It is usual to make the lens-holders capable of being moved slightly out of centre laterally, to allow the lamp to be set so as to show the light at an angle to the signal. This is often required where the line is on a curve. The angle of the cone of light sent out from a lens is not very great (with some lamps about 15 degrees), and the cone must be set to show across the curve to give the driver the best possible view of the light. The lens is adjusted to suit the conditions after the signal has been erected.

GROUND SIGNALS

THERE are three types of ground signals commonly in use:

1st. Discs which partially revolve, showing red, or yellow, and green to the driver as required.

2nd. Discs which are arranged to move downwards, hiding the red or yellow target from the driver when showing clear.

3rd. A miniature semaphore arm, which properly speaking should not be termed a disc, but which serves the same purpose as a ground shunting, signal.

The first class consists of a lamp case fitted with lenses and round targets, showing red or yellow and green as the circumstances may require.

The lamp case can be the same as used for the standard main line semaphore signals, and can be illuminated in the same manner, and with the same interior case and cistern if oil is adopted. The outer case is bolted to a face plate which is capable of moving through 90 degrees, and is operated by a crank and a counterbalance lever. When the counterbalance lever with weight attached is in its normal position, the red (or yellow) lens and target (Fig. 91) faces the driver; when the signal wire is pulled, the lever engages with the crank and moves the lamp round through 90 degrees, and, in so doing, moves the green lens and target to face the driver, the other target, of course, being moved away at the same time. In one variety of this signal, the body of the lamp is fixed immovably, and the targets are fastened to a framing which revolves. This framing carries coloured glasses, there being a common white lens fixed in the stationary case (Fig. 92).

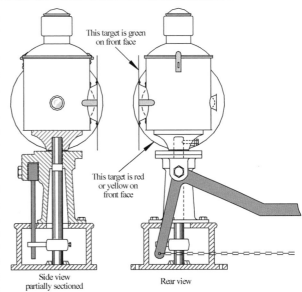

Figure 91
Ground disc signal with moving lamp case

The stand to which the lamp and moving parts are fitted is made of cast iron, with bearings and stops formed to suit the case.

In the 2nd type, the target is fixed to a casting, one end of which is free to move up and down, the opposite end being held in a bearing. This casting is connected by links to the counterbalance lever, so that when normal the target is up and faces the driver, when the signal wire is pulled the casting and target move down and the driver then sees the body of the signal and the edge of the target. The body is made of cast iron, and is fitted with the necessary bearings, etc., for the moving parts. Inside the cast-iron body a small interior lamp (if oil lighting is adopted) is placed.

The lens is carried by the cast-iron body. The red or yellow glass is fitted to the moving target, and the green glass is fitted to a sliding carrier inside the body, which is forced up by the target casting as the target is lowered. When the target is raised to its normal position the green slide falls down below the level of the lens (see Fig. 93).

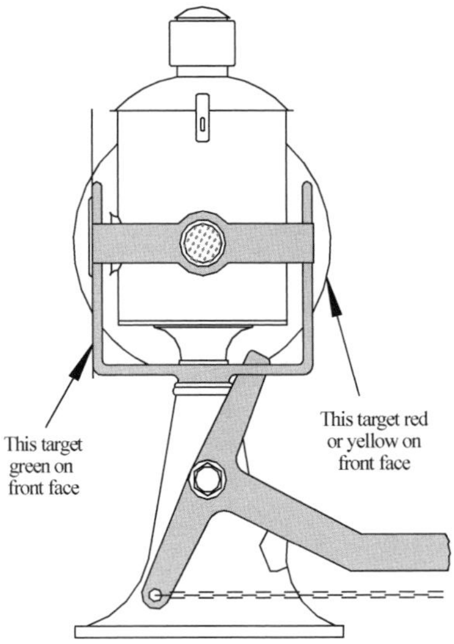

The target is counterbalanced, so that should the pins of the connecting link come out the target will automatically go to the danger position.

The lens of the back light is fitted to the main body casting, and the same slide, which carries the green glass, obscures the back light when the signal shows clear.

The 3rd type consists of a lamp, which may be the standard semaphore lamp, fixed on a casting. This casting is fitted with the necessary bearings and stops for the counterbalance weight lever, and provision made for fixing a small semaphore arm to it. The arm is usually a casting with a small spectacle combined in the same casting (see Fig. 94). Occasionally the portion of the semaphore projecting to the left is made of wood and hinged, so that it can be bent round against the lamp without being damaged. Springs are provided to bring the end of the semaphore back in line with the spectacle again, should it be so bent round by accident.

Figure 92
Ground disc signal with stationary lamp case

The miniature semaphore signal possesses the following good points:

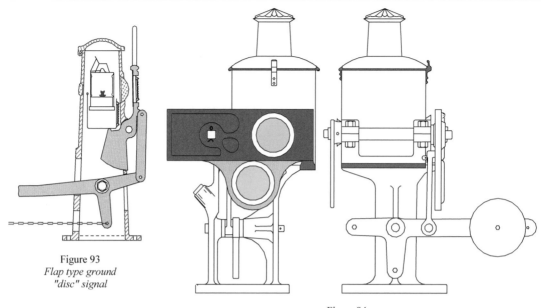

Figure 93
Flap type ground "disc" signal

Figure 94
Ground semaphore signal

1st.   The position of the arm denotes the indication to be conveyed. In the others the colour of the target, or the absence of the target, is the indication for clear, and the colour or presence of the target the indication for danger.

2nd.   Standard pattern lamp cases and interiors can be adopted, and the lamps are stationary.

3rd.   The moving parts are few and of light weight, consequently it is mechanically efficient.

Some of the target signals can lay claim to one or even two of the above items, but no one type can lay claim to them all.

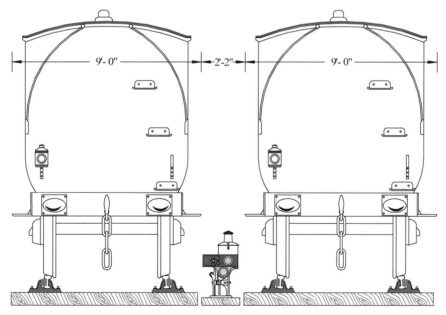

Figure 95
Load Gauge

In all the types of shunting ground-signals the following points must be observed:

- The over-all height of the signal must not exceed 3 ft. above rail level. This is the limit of height for any structure which requires to be fixed in a 6 ft. space between main running lines, and as this is a very common position for disc signals to occupy, it is essential that this rule should be observed (Fig. 95).

- The width of the signal over-all should not be more than 18 in. for the same reason; 18 in. width only allows a clearance of 4 in. between the signal and load gauge on either side. They should be so designed that it is possible to place one signal above another, the combined height not to exceed the 3 ft. above rail level. It is sometimes necessary to have a double signal (the top arm or disc reading to the left) which is capable of being placed in a 6 ft. space.

- The bases should be made so that they can be attached, by means of coach screws, to timbers fixed in the ground; or, if necessary, they should be capable of being fixed on a short post to increase their height.

- The lens should be small, about 2 in. diameter, so as to be distinct from main line signals. All parts should be easily accessible for cleaning and oiling.

WIRES AND PULLEYS

THE wire used for working signals generally is made up of seven strands, the gauge of each strand being 10 or 18 S.W.G.,† and the pitch of the twist about $2^3/_4$ in. to 3 in. A good quality soft steel signal wire should have a breaking load of not less than 2,000 lb.

In laying out wires the run should be as straight as possible; where right-angle bends have to be negotiated, a length of chain should be inserted for passing round the pulley wheel. If the wire itself passes round a pulley it does not last long unless the pulley is of a very large diameter, and the wire well greased. The size of these pulleys varies from 6 in. to 24 in. diameter, the most common size being 10 in. (Fig. 96). Wheels placed near the signal box should be of larger diameter than those placed nearer the signal post, as the travel of the wire is greatest at the signal box. These pulleys, when placed outside the signal box, are usually horizontal, and are screwed down to timbers let into the ground. For leading up to the signal lever in the signal box, vertical or pedestal wheels are employed; these are attached to an upright casting (Fig. 97) which is bolted to timbers. At the signal a swivel pulley, or socket pulley, is used, the swivel being fastened to

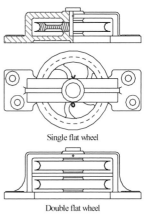

Single flat wheel

Double flat wheel

Figure 96
Horizontal pulley wheels

the base of the post (Fig. 98), which allows it to move round to suit the direction of the run of the wires. The timbers for pulley wheels are called "stools" or "horses," and consist of 4 ft. by 12 in. by 6 in. timbers spiked down to similar timbers which are let into the ground (see Fig. 99). Concrete is often used in lieu of timber.

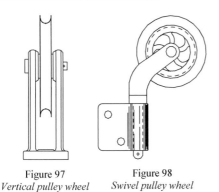

Figure 97
Vertical pulley wheel

Figure 98
Swivel pulley wheel

To support the wires on a straight run small wheels about 2 in. diameter are employed; these are fastened either to wood posts, 3 in. by 3 in., or iron stumps of channel section, $1^1/_4$ in. by $5/_8$ in. by $3/_{16}$ in. (Fig. 100). The pulleys are sometimes galvanised and are generally made of pressed steel or cast iron.

Where a gradual curve requires to be negotiated "swivel," or "crown" pulleys are used; these are fixed to stronger stumps, 4 in. by 3 in., or if made of iron are of T section, 2 in. by $1^1/_2$ in. by $3/_{16}$ in. The length of stumps is about 3 ft. to 4 ft., and they are pitched about 25 ft. apart for a straight run and about 15 ft. on a curve.

† S.W.G. = Standard wire gauge

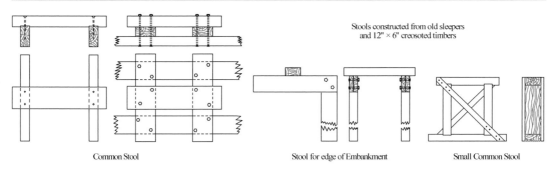

Stools constructed from old sleepers
and 12" × 6" creosoted timbers

Common Stool

Stool for edge of Embankment

Small Common Stool

Figure 99
Timber stools for wheels and cranks

Great care must be taken in setting out a run of wires to see that they will not form an obstruction to shunters and others who have to work along the lines. If the wires cannot be moved out of the way of shunters they must be kept very low down and be covered in with boxing or trunking (see Fig. 152. p. 96). Any wires crossing a path used by platelayers or workmen also have to be boxed in securely.

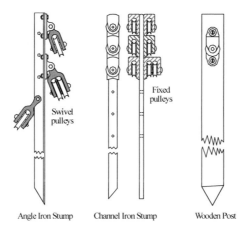

Swivel
pulleys

Fixed
pulleys

Angle Iron Stump

Channel Iron Stump

Wooden Post

Figure 100
Signal wire pulleys and stumps

Joints in the wire and joints between the wire and chain are usually made with eyelets or thimbles round which the wire passes, and are finished off with binding wire.

4
Methods of Working Trains

On double lines, or lines on which traffic is worked in one direction only, the following methods of working are used:

1. Absolute Block.—This applies to all passenger trains (with a few exceptions), and ensures a clear line for a train to a point at least 400 yards (preferably 1/4 mile) past the home signal, before it is accepted from the rear signal box.

2. Block Working under Warning Arrangements.—This applies to non-passenger trains running on a line which is equipped for absolute block working. (Under exceptional circumstances passenger trains are allowed to travel under this arrangement.) A train may be accepted when the line is clear only to the home signal, but the driver must be warned of the fact at the rear box.

3. Lock and Block.—This interlocks the signals for entering the section with the block instruments making it impossible for the signalman to forward a train without the permission of the signalman at the box ahead. It is used on very busy lines, or on lines which are difficult to work, such as underground lines, where the signalman has an imperfect view of trains.

4. Time Interval or Recording Telegraph System.—This applies to non-passenger trains only, and on lines set apart for that purpose; it allows more than one train in the section between two boxes at the same time.

Block Telegraph Working (Double Lines)

The block telegraph system is for the purpose of providing an adequate interval of space between following trains, and in the case of junctions between converging or crossing trains, as required by the Ministry of Transport.

Given a portion of a railway to equip with instruments, etc., for block working, it is first necessary to divide the railway into sections, each section being termed a "Block Section". At the end of each section a signal box equipped with block telegraph instruments and also outdoor signals must be provided.

Block signal boxes must be situated at all places where trains can cross or in any way foul the path of other trains. Hence, all junctions between running lines, whether running junctions or cross-over roads, must be controlled from a block signal box. It is not, however, *necessary* to have a block signal box at all stations, nor at sidings which connect with *one running line only*.

A passenger train must not be allowed to enter a block section unless there is no other train occupying that section at the time, and after the first train has entered the section no other train must be allowed to foul the section in any manner until the first train has passed out of the section.

The time between successive trains is regulated by the length of time taken to run through the section; hence it follows that the carrying capacity of the line depends to a great extent on the distance between block signal boxes. On busy lines it is economical to make each station, or intermediate siding, the end of a block section, and, in some cases, to cut up the line between two stations into two or more block sections by erecting a box for block signalling purposes only, such a box being termed a Block Box. On lines which are not busy it is economical to have the sections of some considerable length, in which case a station might be in the middle of a block section.

Where a block box is equipped with starting signals, the block section is regarded as commencing at the Starting (or Advanced Starting) signal, and terminating at the Home signal of the box ahead, the interval between the Home and Starting signals being termed the Station Limits.

The portion of line which forms the BLOCK SECTION must be regarded purely as a RUNNING LINE, and that *only* in the predetermined direction. If a train breaks down *in* the section the guard must take steps to protect his train from being run into by a following train (unless protected by fixed signals), and not rely on the protection of the block working arrangements. Trains must only be allowed to move in the wrong direction in the block section after the engine driver has received written permission to do so. Within station limits a train may

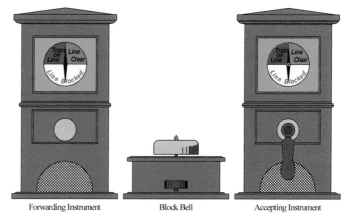

Forwarding Instrument   Block Bell   Accepting Instrument

Figure 101
Needle type block instruments

be left standing, being under the protection of fixed signals, and trains can be moved in any direction under the verbal instructions of the signalman.

In practice, PASSENGER TRAINS are not allowed to enter a section unless the section is unoccupied, *and, in addition,* the line ahead must also be *unoccupied* for a distance *not less than* 400 yards. (Most railway companies adopt $1/4$ mile = 440 yards as the limit.)

Fig. 101 shows a common type of block instrument used for double lines. Two instruments are required for each section, one with an operating handle being fixed in the signal box at the "leaving" end of the section, and an instrument, without an operating handle, is fixed at the "entering" end of the section. The instruments in the two signal boxes are connected electrically, so that a movement of the operating handle moves the needles of both instruments simultaneously, each giving the same indications. The instrument with the handle is for the purpose of ACCEPTING trains from the box in the rear, and the one without the handle is to tell the signalman when the train has been accepted by the box ahead. The instruments must at all times give a faithful record of the state of the section *plus* the over-run; when the section is unoccupied, and permission has NOT been given for a train to enter the section, the needles hang vertically—this indicates LINE BLOCKED; when permission is given to forward a train the needle is moved to the side of the dial lettered LINE CLEAR; and, finally, when a train is in the section, or if for any reason the section, *plus* a quarter of a mile beyond the end of the section, is not clear, the needle must be pointing to the portion of the dial lettered TRAIN ON LINE.

A bell of the single beat pattern is installed for each section in both boxes, the bells being electrically connected. The function of the bell is, in the first instance, to CALL ATTENTION, and, secondly, to indicate by means of a code of rings information as to the class of train being dealt with, etc., etc. The codes vary very slightly in minor details on the various railway companies; the code used by one railway is given on the opposite page.

In addition to the block instruments, each signal box is supplied with a train register in which the time of receiving all block signals must be recorded; also information as to the class of train, the station from which the train started, and its destination must be given. The register is used as a journal, and any unusual occurrence must be recorded therein.

Method of forwarding trains, absolute block system: Assume two signal boxes "A" and "B" (see Fig. 102).

If the signalman at "A" has a train to forward to "B", the following routine must be carried out.

It is assumed that all the outdoor signals and block instruments are in the normal position.

The block instrument for the line from "A" to "B" shows LINE BLOCKED; this simply means that the line is CLOSED.

(Continued on page 58)

Bell Signals

	Beats on Bell	
Call Attention	1	1
Is Line Clear for Express Passenger Train, or Break-down Van Train going to clear the Line, or Light Engine going to assist disabled Train?	4	4 consecutively
Ordinary or Excursion Passenger Train, or Break-down Van Train NOT going to clear the Line?	4	3 pause 1
Branch Passenger Train? (Applicable only where special instructions are given)	4	1 pause 3
Fish, Meat, Fruit, Horse, Cattle, or Perishable Train, composed of Coaching Stock?	5	5 consecutively
Empty Coaching Stock Train?	5	2 pause 2 pause 1
Fish, Meat, or Fruit Train, composed of Goods Stock ; Express Cattle or Express Goods Train?	5	1 pause 4
Through Goods, Mineral or Ballast Train?	5	4 pause 1
Ordinary Goods, or Mineral Train stopping at intermediate Stations?	3	3 consecutively
Branch Goods, Mineral or Ballast Train? (Applicable only where special instructions are given)	3	1 pause 2
Through Goods, Mineral or Ballast Train?	5	4 pause 1
Light Engine, or two Light Engines coupled, or Engine and not more than two Brakes?	5	2 pause 3
Ballast Train, or other Train requiring to stop in Section?	5	1 pause 2 pause 2
Platelayers' Lorry requiring to pass through Tunnel?	5	2 pause 1 pause 2
a Train entering Section	2	2 consecutively
a Assistant Engine in rear of Train	4	2 pause 2
a Obstruction removed or Train out of Section	3	2 pause 1
Obstruction Danger	6	6 consecutively
Blocking Back	6	Inside Home Signal—2 pause 4 / Outside Home Signal—3 pause 3
Stop and Examine Train	7	7 consecutively
Cancelling Signal	8	3 pause 5
Caution Signal	7	4 pause 3
Train passed without Tail Lamp	9	9 consecutively to cabin in advance / 4 pause 5 to cabin in rear
Train Divided	10	5 pause 5
Shunt Train for following Train to pass	11	1 pause 5 pause 5
Vehicles running away on wrong line	12	2 pause 5 pause 5
a Section Clear but Station or Junction blocked	13	3 pause 5 pause 5
Vehicles running away on right line	14	4 pause 5 pause 5
Opening of Signal Cabin	15	5 pause 5 pause 5
Testing Block Indicators and Bells	16	16 consecutively
Closing of Signal Cabin	17	7 pause 5 pause 5
a Time Signal	18	8 pause 5 pause 5
Lampman or Fog-signalman required	19	9 pause 5 pause 5
Testing Controlled or Slotted Signals	20	5 pause 5 pause 5 pause 5

a CALL ATTENTION—The Call Attention signal must always be given before any other signal, except those marked a, and must be acknowledged immediately on receipt.

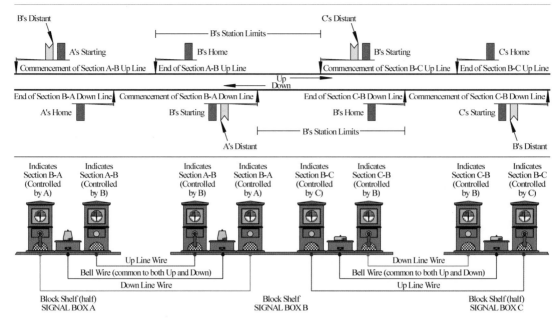

Figure 102
Block working with needle instruments

(Continued from page 56)

- If the last train has been passed through the section in the ordinary manner, "A" must first give the ATTENTION signal on his bell key. "B" will return acknowledgment by repeating the ring.

- "A" then gives the "IS LINE CLEAR?" signal applicable to the class of train to be sent. "B" must satisfy himself that he can safely give permission for the train to approach. If the train is a passenger train the section "A—B" must be clear, and, in addition, the line ahead of the home signal for a distance of not less than 440 yards. This distance is termed the "OVER-RUN", as it allows a clear line in the event of a driver being unable to bring his train to a stand at the home signal. The rule also requires that if a train should happen to have just cleared the over run it must be *proceeding on its journey* before the following train can be accepted; and should the train leaving "B" break down just ahead of the clearing point, the second train must not be accepted. If the section and the over-run are both clear, "B" will acknowledge the "Is Line Clear?" signal by repeating it, thus letting "A" know that his signal has been correctly received by "B", and then "B" must move the handle of his block instrument for the section "A—B" to the LINE CLEAR position. This causes the needles in both boxes to move to the LINE CLEAR position. "A", on seeing his instrument show LINE CLEAR, can lower his starting signal for the train to proceed to "B".

- As the train leaves "A" the signalman there must give the rings on his bell indicating TRAIN ENTERING SECTION; "B", on hearing this signal, must acknowledge the signal by repeating it and move the handle of his block instrument for the section "A—B" to TRAIN ON LINE. This moves the corresponding needles to TRAIN ON LINE in both boxes. When the train has reached "B" with tail light complete, and passed out of the section, past the over-run,† and is proceeding on its journey, the signalman must call "A's" attention; "A" must acknowledge by repeating the signal: "B" then gives the rings for TRAIN OUT OF SECTION, and places the handle of the block instrument in its normal position, "A" acknowledging the signal by repetition.

---

† Some railway companies' rules instruct the signalman to give Train Out of Section to the box in the rear as soon as the last vehicle of the train with tail light has passed the home signal. This is undesirable, as although the section itself is clear, the over-run is occupied, and permission could not be given for a second train to approach from the box in the rear.

It will, of course, be understood that the signalman at "B" can give the TRAIN OUT OF SECTION signal to "A" if he has shunted the train clear of the running line (assuming he has a siding or cross-over road), instead of forwarding it on to the next box.

After a signalman has given permission for a train to approach towards his box from the box in the rear, he must not allow any obstruction to foul the running line on which the train has been accepted.

Before the line is considered clear, all points over which the train about to be accepted will run must be in the correct position for that train.

A signalman *must not ask the box ahead to accept* a passenger train unless he has received TRAIN OUT OF SECTION for the preceding train, and the block indicator is in the normal position.

If a signalman shunts a train from one running line to another, the over-run of the line on which the train has been shunted is no longer clear; therefore, the permission of the box in the rear of the section concerned must be obtained before the train can be shunted; the BLOCKING BACK INSIDE HOME SIGNAL must be given and acknowledged, and the block instrument for that line must be made to indicate TRAIN ON LINE before the shunting movement is made.

On some railways the code gives separate rings for a passenger and non-passenger train, as it is necessary for the rear box to know what description of train is occupying the over-run; for example, should it be a passenger train, he would be prohibited from offering another train of any description on that line. If a train is moved out of a siding into the over-run, the same routine must be observed, unless the train is going ahead into the next section *without delay*.

If it is desired to shunt a train outside the home signal into the rear section, the permission of the signalman in the rear must first be obtained by giving BLOCKING BACK OUTSIDE HOME SIGNAL, and the instruments concerned must show TRAIN ON LINE, the code of rings again giving the information as to whether the train is a passenger or non-passenger one. The signalman receiving the blocking back signal must see that he can safely give permission for the section or the over-run to be occupied.

Where the block boxes are so close together that the over-run of one box reaches to the home signal of the box ahead, the rear box must not accept a train until the box ahead has first accepted it; similarly, the blocking back signal must not be accepted by the box ahead until the rear box has accepted it.

Fig. 103 shows the signalling arrangements for a junction with the adjacent signal boxes. The signalman at "B" must not accept two trains which can cross the path of each other at the same time.

Thus he cannot accept a train from "A" to "D" at the same time as he has accepted a train from "C" to "A", nor must he accept a train from "C" and another one from "D" at the same time.

By setting his facing points, so as to secure the required over-run, he can, however, accept a train from "A" destined to go to "D", at the same time that he has accepted a train from "C" to "A". He can safely do so by keeping his signals for the train from "A" at danger, and having his facing points set for "C", so that in the event of the train over-running, it would proceed towards "C", and not cross the path of the train from "C" to "A". If there is already a train standing at the starting signal of the line towards "C", and a second train is required to be accepted from "A" to go to "C", the second train can be accepted if the facing points are set for the train to run to "D" in the event of its over-running the home signals, and no train has been accepted from "C" to "A" at the same time.

This ruling only applies to running trains, and a train may be accepted from "A" to go to "D", if the train accepted from "C" has come to a *stand* at the home signal clear of the crossing. Similarly, a train may be accepted from "C" to go to "A" if the train from "D" has come to a stand at the home signal, clear of the fouling point.

The chief essential is that there shall be *one clear section plus the over-run before a passenger train can be accepted,* and it is of no importance how the over-run is obtained; the provision of an outer home signal 440 yards in the rear of the inner home signals at junctions such as "B" is quite sufficient; this, however, ignores the provision that the train must not only be 440 yards clear, but also *proceeding on its journey*. The section "C—B" terminates at the outer home signal, and the over-run terminates clear of the fouling point of the junction; similarly, the section "D—B" terminates at the outer home signal on that line, and the corresponding over-run terminates well clear

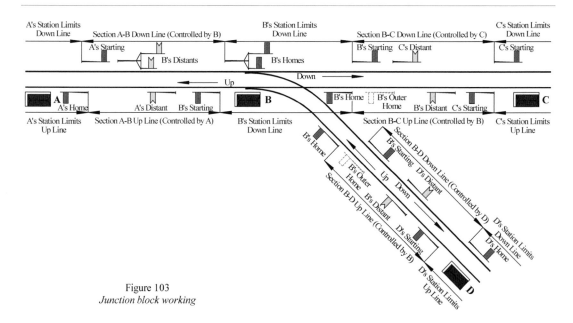

Figure 103
Junction block working

of the fouling point of the junction, therefore in a case of this sort trains may be accepted simultaneously from "C" and "D".

BLOCK WORKING UNDER WARNING ARRANGEMENT

THIS is termed the WARNING ARRANGEMENT, because the signalman forwarding a train under this regulation must warn the driver that the line is only clear to the HOME signal. Working under this arrangement is often termed RULE 5 WORKING, because that is the number of the rule in the Standard Railway Clearing House Regulations for Block Working.

Where the line is clear only to the home signal, the over-run being occupied, and it is desired to accept a second train, then, if the first and second trains are *both* NON-PASSENGER trains, the second train may be offered and accepted. The signalman accepting this train does not repeat the "Is Line Clear?" signal, but gives SECTION CLEAR BUT STATION OR JUNCTION BLOCKED signal in its place; the signalman receiving this signal warns the driver that the line is only clear to the home signal. The needles of the block instruments concerned indicate TRAIN ON LINE, all the time that the section or over-run is occupied, and TRAIN OUT OF SECTION must not be given. until both the section and the over-run are clear by both trains having passed out of the section. Trains can only be dealt with under this rule in clear weather. By obtaining exemption from absolute block working, passenger trains may be accepted under the "Warning Arrangement".

On most lines during some portion of the twenty-four hours traffic is slacker than at any other portion of the day or night. If the slack period is of several hours' duration it is convenient to be able to work with longer block sections. To enable this to be done SWITCHES are provided in some of the signal boxes, allowing the box to be closed and the block instruments of the boxes on either side to be connected up. Thus, if A, B and C are three boxes, B can be "SWITCHED OUT" and the instruments at A and C are connected together instead of A with B and B with C.

On certain lines for Sunday working, a large number of boxes are closed and the block section may be many miles long, with all the intermediate boxes switched out. Before a box is switched out by the signalman all the running signals for the line concerned must be pulled to *clear*. The box is no longer a block box, and at all boxes which are not block boxes the signals must be in the clear position, unless required to be put to danger to protect a train or obstruction.

The signalman is not allowed to switch out until the sections on either side of him are unoccupied by trains. On switching in the signalman must be informed of the position of the trains (if any) in the sections controlled by him, and of any special circumstances which may require his attention.

With the needle type in use the instrument without the operating handle is included in the circuit; this informs the signalman on arrival of the state of the sections before he switches in.

LOCK AND BLOCK WORKING

WITH the ordinary block signal arrangements there is nothing to prevent the signalman from pulling his starting signal to CLEAR before he receives permission from the box ahead.

There are several instruments designed to prevent this possibility.

This method of working is termed LOCK AND BLOCK working. Fig. 104 shows the face of a SYKES' Lock and Block instrument. The connections to the starting signal lever are similar to the arrangement shown in Fig. 213, p. 137. Where boxes are fitted with this instrument, the following mode of working is adopted:

The usual code of bell signals is employed, and given on an independent bell. If the line is clear the signalman at box "B" presses in his plunger; this sends an electric current to the instrument at Box "A", and releases the lock on the starting signal. The instrument at "A" indicates FREE, informing the signalman that he can pull his starting signal lever. At "B" the lower indicator shows

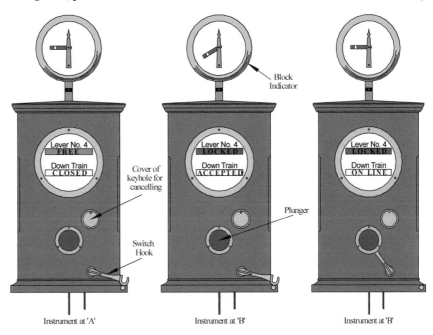

Figure 104
Sykes' lock and block instrument

TRAIN ACCEPTED. The plunger at "B" having once been pressed in to release the starting signal at "A", cannot be depressed a second time until the train accepted has passed through the section. When the train leaves "A", the usual TRAIN ENTERING SECTION signal is given on the bell; "B" acknowledges it by repetition, and at the same time turns his switch hook over his plunger; this causes the small semaphore arm on the top of the instrument at both boxes to go to the danger position, indicating that the section is occupied, and at the same time the TRAIN ACCEPTED indication at "B" is covered by a disc which shows TRAIN ON LINE.

The starting signal lever at "A", when once it has been pulled to allow a train into the section, is back-locked until the train has passed over a treadle a train's length past the signal, when the signal lever can be restored to the normal position. The back-locking is arranged in most cases so that the lever can be put sufficiently back to place the signal to danger, if required, in the case of emergency, but not far enough back to operate, or release, any interlocking. When the train arrives at "B" and has passed over the treadle ahead of the starting signal (the box at "C" having accepted the train), the starting signal at "B" can be restored to the normal position, and the replacing of the starting signal lever re-sets the instrument to its normal position. A second train can then be accepted from "A".

In most installations the starting signal is fitted with a "reverser", so that the train passing the treadle ahead automatically places the signal to danger behind it, and the signal cannot be lowered until the starting signal lever has been put back, and then pulled again, but the lever cannot be pulled until the lock has been released by the signalman ahead. Any movement of the starting signal lever causes it to be locked either in the normal or in the reversed position. The normal lock can only be released by the signalman ahead, and the reverse lock can only be released by a train passing over the treadle ahead of the starting signal (in the ordinary course of events).

At stations where there are sidings into which the train can be shunted clear, or at junctions where a train is brought to a stand at a starting signal, and the facing points moved to run the following train on to a different line, additional treadle contacts must be provided to allow the signalman to accept a second train as soon as a CLEAR section plus an over-run has been provided.

In the case of shunting the first train into a siding to clear the main line for a second train, treadles can be fixed on the siding to give the necessary clearing of the instruments, and in the case of junction working an electric contact on the facing point lever can be used to give the necessary relief.

On lines where the SECTION CLEAR BUT STATION OR JUNCTION BLOCKED ruling is in force, the clearing treadles can be placed immediately clear of the home signal, and on this signal being pulled over and then restored to DANGER the instrument must be re-set to allow a second train to be accepted. The home signal in this case is locked in the reversed position until the train passes over the treadle. The arrangements for locking and releasing the starting signal will be similar to the case in which that signal lever on being put back re-sets the instruments. It will not, however, be locked in the reverse position, as that is done by the home signal.

A better system for allowing the SECTION CLEAR BUT STATION OR JUNCTION BLOCKED arrangement to be used with lock and block working, is to provide a special arm below the starting signal. The lowering of this arm indicates to the driver that absolute block working is not in force and that he must be prepared to stop clear of the next home signal. The "warning" arm is worked by a separate lever which is not locked with the block instrument.

The block telegraph rules require all points over which the train has to pass to be in the correct position before a train can be accepted, and as no obstruction to the line on which the train is about to travel can be allowed after the train has been accepted, until the train has either come to a stand at the home or starting signal, or has cleared the section, the points connected with that line should not be moved to allow any vehicles to foul the line concerned.

At through-running junctions all the necessary safety as to crossing trains, holding and locking points, etc., can be ensured by arranging that until the home signal has been pulled, a train cannot be accepted on the instrument from the box in the rear. The home signal, having been pulled over to release the block instrument plunger, is locked in the reversed position until the train arrives at, and passes over, a treadle fixed AHEAD of the home signal.

At junctions which are stations it would be very inconvenient to compel the signalman to lower the home signal before accepting a train, and to obviate this a special setting lever can be introduced. This lever does all the necessary inter-locking of points, and it must be pulled before a train can be accepted on the instrument, and once it has been pulled and a train accepted it cannot be put back to unlock the points until the train has arrived at a treadle, or balanced bar, fixed immediately in the REAR of the home signal. This allows a train to come to stand at the junction home signals, and the points can then be moved as may be required for traffic purposes. When the treadle releasing the setting lever is in the REAR of the home signal, the block instruments must not be released to accept a second train from the rear, until the first train has passed over a second treadle AHEAD of the home or starting signal, as may be required. The setting lever can also be used to advantage at stations other than junctions, where it is necessary to shunt a train from one running line to another, as the pulling over of the cross-over road points will be made to lock the setting lever, and so prevent a train from being accepted on the occupied line; conversely, if a train has been accepted on the other line, the setting lever being pulled prevents the cross-over road lever from being used.

By the use of a treadle and a fouling bar in conjunction, it can be arranged that, until the whole of the train has passed over the treadle and bar, the required release is not given. This is chiefly necessary when the treadle is on a siding for the purpose of clearing the section after the first train has been shunted into the siding off the main line.

Lock and block working can be carried out in conjunction with almost any type of block instrument, but with all types of instruments some arrangement has to be made for cancelling or resetting the instrument in case of failure of the treadles or plungers, and also in the event of a train being cancelled after having been accepted. It is now usual to compel the signalman to obtain permission from the signalmen on either side of him before he can obtain the desired release. With SYKES' instrument this is done by fitting an electric lock on the cover of the keyhole in the instrument, so that the two men on either side must press a button before the signalman desiring the freedom can insert a re-setting key. With the TYER'S pattern the re-setting is effected by means of a relieving instrument. This is fitted with a disc handle which can be rotated, and is lettered: (1) NORMAL; (2) RELIEF TO BOX "A"; (3) RELIEF TO BOX "C"; (4) KEY.

To release the instrument the signalmen on either side must move the discs in the direction giving relief to the box asking for that relief; the signalman at the latter box then turns his disc to KEY, and can thus gain access to the top portion of the instrument. By manipulating the required switches either "LEVER LOCK" or "INDICATOR AND KEY LOCK" he can then carry out the required re-setting or releasing. These arrangements involve three signalmen, which is a guarantee against the relief being obtained for an irregular purpose.

TRACK CIRCUIT BLOCK

PARTLY because of the weakness introduced into the LOCK AND BLOCK system by the CANCELLING or RE-SETTING feature, and partly because of the complicated nature of the instruments, treadles, etc., there has recently been a tendency for Lock and Block to be replaced by "TRACK CIRCUIT BLOCK".

With TRACK CIRCUIT BLOCK the lines are completely track-circuited from the STARTING (or ADVANCED STARTING) signal of one box to the end of the over-run or "overlap" at the next box. An electric lock (see p. 135), controlled by this track circuit, is provided on the lever operating the STARTING (or ADVANCED STARTING) signal, which can then only be cleared when the track circuit is "clear". The usual method of working is then for the signalman at (say) "A" box to pull off his STARTING (or ADVANCED STARTING) signal at any time, provided the track circuit is "Clear" without asking acceptance from "B" box, and to give to "B" a special "ENTERING SECTION" code ring on a bell, in order to advise "B" of the approach of a train and its description.

TIME INTERVAL OR RECORDING TELEGRAPH WORKING

ON lines used exclusively for Goods Traffic, trains may be allowed to enter the section when not only is the over-run occupied, but the section itself is also occupied; the trains following may be sent after a certain interval of time, or on slow-speed lines no interval may be prescribed. Instruments may or may not be used, but a bell is generally employed for signalling purposes.

Fig. 105 shows an instrument used on lines worked under the Recording Telegraph arrangement.

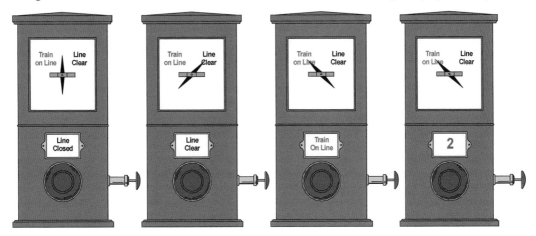

Figure 105
Recording block instrument

Instead of a long handle, as in the case of the ordinary needle instrument, a circular handle is provided. Normally, the needle hangs vertically, the indicator then reading LINE CLOSED. To accept a train, the handle is turned slightly and the needle moves to the LINE CLEAR position, the indicator showing the same reading. When the train enters the section the handle is turned further round, the needle then takes up the TRAIN ON LINE position, the reading of the indicator corresponds; if a second train is admitted into the section before the first is clear, the handle is turned still further, the needle remains at TRAIN ON LINE, but the indicator reads 2, to signify that two trains are in the section. The instruments are designed to accommodate about six trains. As a train passes clear of the Home signal, "TRAIN OUT OF SECTION" is given to the box in the rear, and the handle turned back one step, thus indicating one train less in the section. The side plunger is for the purpose of preventing the handle from being moved more than one step at a time, as after each step the handle becomes locked until released by pressing in the side plunger.

These instruments can be used for either ABSOLUTE BLOCK or WARNING arrangement working, in addition to purely non-passenger line work. In accepting non-passenger trains, in good weather, if the line is clear to the home signal they are accepted at LINE CLEAR by repeating the IS LINE CLEAR? signal, but during foggy weather, when the over-run is occupied, and the section itself clear, SECTION CLEAR BUT STATION OR JUNCTION BLOCKED signal is used; when the SECTION is occupied, a special signal CAUTION (4 pause 3) is given. The instrument at the receiving end always gives a faithful record of the state of the section. At the sending end, however, there is nothing except the ordinary needle indicating the usual "LINE BLOCKED", "LINE CLEAR" and "TRAIN ON LINE"; the signalman sending the trains has to refer to his train register book if he wishes to know how many *trains are in the section ahead.*

Where a line not worked on the ABSOLUTE block system joins another line so worked, the signalman at the junction signal box must only accept trains on the converging lines under ABSOLUTE BLOCK Regulations, unless safety-points are provided on the non-absolute block line, when, so long as the safety-points are set so as to protect the absolute block line, trains may be dealt with on the non-passenger line under non-absolute block regulations.

SINGLE LINE WORKING

ON railways where traffic runs in both directions on the same line, the methods of working are given in Part C. of the Ministry of Transport Requirements (see p. 70).

Fig. 106 shows one pattern of single line staff.

1. *Train Staff and Tickets.*—The line is equipped with block instruments of a similar type to those used for double lines, and the regulations for working trains are similar, with the following exception as to accepting trains at LINE CLEAR.

At a train staff station, which is a passing place, LINE CLEAR may be given if the line, on which the approaching train has to run, is clear to the starting signal, and the facing points set for that line. It is possible that the starting signal may be considerably less than 440 yards from the home signal, so that the distance beyond the home signal for an over-running train may be *less* than the standard 440 yards. Where special over-run sidings are laid down this exception disappears.

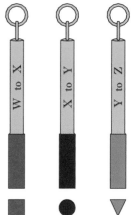

Figure 106
Single line train staffs

Rule as to working signals at PASSING PLACES:

"When trains which have to pass each other are approaching a staff station in opposite directions, the signals in both directions must be kept at DANGER, and when the train which has first to be admitted into the station has come to a stand, or where there are over-run sidings, nearly to a stand, the home signal applicable to such train may be lowered to allow it to draw forward to the station or the starting signal, and after it has again come to a stand, and the signalman has seen that the line on which the other train will arrive is quite clear, the necessary signals for that train may also be lowered".

The fundamental principle on which STAFF AND TICKET working depends is that a metal bar or staff (see Fig. 106) is provided for each stretch of single line between passing places, and very rigid instructions are given to engine drivers that no engine or train must proceed on to the single line (except under certain conditions, for shunting purposes) unless the driver has the staff in his possession or has it shown to him before he starts.

There is only one staff for each single line section (between passing places), and it has engraved on it the name of the section to which it belongs, and, in order still further to reduce the risk of the staff for one section being mistaken for that of another, the staffs for adjoining sections are made of different shapes and colours, as, for example:

Section W to X:
Square   Coloured Red.
Section X to Y:
Round   Coloured Blue.
Section Y to Z:
Triangular   Coloured Green.

The possession of the staff by every train running over the section would ensure safety, but would be a great hindrance to traffic, as it would necessitate trains running alternately in each direction, whereas the requirements of traffic might call for more trains in one direction than the other. To meet this demand the Train Tickets are brought into service, and the instructions to drivers permit any number of trains to follow each other, each driver having the Staff shown to him and a Ticket given to him, thus authorising him to run on the single line. It must be noted that a Ticket is no authority for this unless the Staff is shown at the time the Ticket is given.

The last train through the section in any direction carries the Staff which, on its arrival at the other end of the section, is the authority for trains to be despatched in the opposite direction.

This method of using a Staff and Tickets provides protection against trains *meeting* in the section, while the ordinary block telegraph pro-

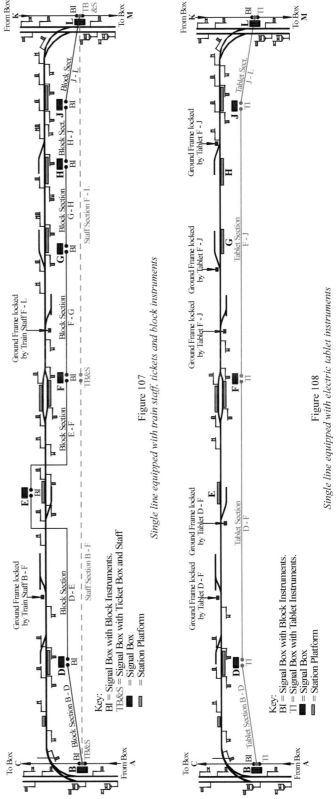

Figure 107

Single line equipped with train staff, tickets and block instruments

Figure 108

Single line equipped with electric tablet instruments

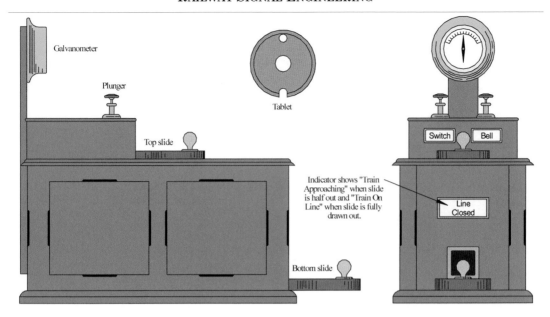

Figure 109
Tyer's tablet instrument

vides protection against trains *overtaking*. It can thus be seen how one Staff section may be divided into two or more Block sections so that two or more trains may *follow* each other at the same time through the Staff section.

In dividing the line into block sections for this mode of working, a block box must be placed at each passing place, sidings between stations being, as a rule, controlled by the train staff, in which case signals are not required.

Stations which are not used as passing places may be fitted as block posts for following trains only, in which case the block instruments are usually kept in the station-master's office, a small locking frame being erected on the platform (with or without cover) for working the points (if any) and signals.

Each PASSING PLACE must be equipped with signals and block instruments, and must be a STAFF STATION. All block posts must be equipped with block instruments and signals.

Fig. 107 shows a line signalled to suit this system.

2. *With one Engine in Steam carrying the Staff on a Single Line.*—With this system neither block instruments nor signals are required on the single line. All points are locked with the train staff. If the single line joins another line, single or double, a block box with signals, etc., will be required at the junction of the lines, and an outer signal, or over-run, must be provided, so that, so long as an engine or train is in the single-engine-in-steam section, a clear over-run of 440 yards will be available for it.

3. *Electric Tablet or Electric Train Staff.*—With these systems ordinary block instruments are dispensed with and outdoor signals are only provided at STAFF OR TABLET STATIONS. All intermediate sidings are controlled by the tablet or staff, and any intermediate stations which may be between Staff or Tablet Stations are *not* equipped with signals of any description.

Staff or tablet stations may exist at places which are not passing places, but all passing places must be equipped as staff or tablet stations (see Fig. 108).

Fig. 109 shows one pattern of Tyer's tablet instruments. There are about twelve tablets in each instrument, and there are two instruments to each tablet section, one at each end of the section. The two instruments are connected electrically, and the arrangements are such that only one tablet at any one time can be out for any one section. If a tablet is taken out of the instrument at "A", the instrument at "A" becomes locked, so that a second tablet cannot be withdrawn, and the instrument at "B" is also locked, so that no tablet can be obtained from that instrument. The method of obtaining a tablet, assuming both instruments normal with the slides in, is as follows (the general block signalling rules are carried out as in the case of system 1):

"A" calls "B's" attention by depressing the plunger marked "BELL". "B" replies. "A" then gives the IS LINE CLEAR signal according to the class of train to be forwarded. "B" acknowledges, and then "A" again holds down his BELL plunger for a few seconds. Whilst the bell plunger is depressed the needle above the instrument is deflected to one side. "B" on seeing the needle deflected, depresses his SWITCH plunger with one hand, and withdraws his bottom slide with the other hand, the slide only coming half way out; he then ceases to hold down the SWITCH plunger, but depresses the BELL plunger. "A" then depresses his SWITCH plunger, and at the same time withdraws his bottom slide right out, and then can obtain the tablet. Having taken the tablet off the bottom slide, the signalman at "A" puts it into a carrier pouch and hands it to the driver. The indications given by the instrument are: bottom slide right in—LINE CLOSED, slide half out (tablet cannot be extracted)—TRAIN APPROACHING, and the bottom slide right out (tablet can be obtained)—TRAIN ON LINE.

The bottom slide having been pulled out cannot be put back again, unless the tablet has been restored to the same instrument by pulling out the top slide, inserting tablet thereon and pushing the top slide in (this returns the tablet to the magazine of the instrument), or unless the tablet has been taken through the section and inserted into the instrument at "B" by means of the top slide. When the tablet is restored to the instrument from which it was obtained (say box "A"), the bottom slide can be put back, and to enable the signalman at "B" to put in his slide (which was pulled half-way out), "A" must give the cancelling or shunting completed signal, and hold down the bell plunger on the last beat. "B" must then depress his switch plunger, when he will be able to return the bottom slide to its normal position. If the tablet is taken through the section to "B",

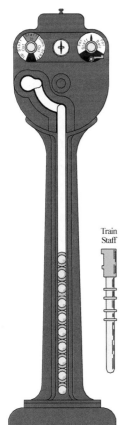

Train
Staff

Figure 110
Electric train staff instrument

on the tablet being restored to the instrument there (by means of the top slide) the bottom slide can be put back to its normal position. "B" then gives "Train out of Section", holding down his bell plunger on the last beat. "A" then must depress his switch plunger, at the same time returning his bottom slide to its normal position.

In order to move the bottom slides, in either direction, the tablet is required to be inserted by means of the top slide, or the signalman at the other end of the section must depress his bell plunger with his bottom slide right in, and the switch plunger of the instrument requiring the movement must also be depressed.

THE ELECTRIC TRAIN STAFF is identical in principle to the tablet system. One staff only can be out for any one section at once; one may be obtained from either instrument, but only by the consent of the other signalman, and the staff can be returned either to the instrument from which it was obtained or to the instrument at the other end of the section.

There are about twenty staffs for each section (some instruments have twenty per instrument). Fig. 110 is a diagram of the instrument as patented by Messrs. Webb and Thompson, of the L. & N. W. R., in 1889.

To obtain a staff, the usual CALL ATTENTION and IS LINE CLEAR Signals are exchanged. "B" must hold down his bell key on giving the last beat acknowledging the IS LINE CLEAR signal. "A" seeing the needle deflected, turns his right hand indicator to FOR STAFF, and he can then obtain a staff from the instrument. On withdrawing the staff the indicator moves from FOR STAFF to FOR BELL automatically. "A" must then turn his left-hand indicator to UP (or DOWN), STAFF OUT, and press it hard down for a few seconds; this causes the needle to assume a vertical position in both boxes, and "B" then knows that "A" has withdrawn a staff and ceases to hold down the bell key.

"B" must then turn his indicator to UP (or DOWN), STAFF OUT. If the staff is taken through the section the signalman at "B" will insert it in his instrument and give the TRAIN OUT OF SECTION signal to "A". Both signalmen then restore their indicators to STAFF IN. The instruments are now in the normal position ready for another staff to be withdrawn at either end.

If the staff is restored to the instrument from which it was obtained, the cancelling or shunting completed signal would be given to the other box, and then both signalmen must return their indicators to STAFF IN, thus placing everything in the normal position.

It will be noticed that the staff last inserted in the instrument is the first to be taken out, which results in a few of the staffs receiving more use and wear than the others. In an improved design a magazine is provided to allow the staffs to be used in rotation. Fig. 111 is a diagram of a miniature staff instrument manufactured by the Railway Signal Co., Ltd., with such a magazine. In the case of the large staff a key for unlocking and locking intermediate sidings is provided on the staff, but with the miniature staffs, the whole staff is inserted in the siding lock in the same manner as a tablet. The rings on the staff are varied in position for each section to prevent the staffs from operating the instruments of sections to which they do not belong.

At staff or tablet stations which are Not passing places, the signalman must on no account accept trains from opposite directions at the same time. If there should be two staff or tablet stations between passing places, "A" being a passing place, "B" and "C" staff or tablet stations but not passing places, and "D" a passing place, then for a train to pass from "A" to "D", "B" must not permit "A" to withdraw a staff or tablet until he has, himself, obtained one for the section "B—C." "C" will not allow "B" to obtain this staff or tablet if he has allowed a train to approach from "D." This, then, prevents two trains leaving "A" and "D" simultaneously. A similar arrangement being in force for trains leaving "D" to go to "A," "C" cannot give "D" permission to withdraw a staff or tablet until he has first obtained a tablet with the permission of "B."

Figure 111
Miniature electric train staff instrument with magazine

It is usual to make a ruling that no train (or engine) must foul the single line unless the driver is in possession of the STAFF or TABLET for that section of the single line, and for this reason, as a rule, no advanced starting signals are provided ON THE SINGLE LINE. Where shunting is required to be performed at both ends of the section simultaneously, a special SHUNTING STAFF is provided at each box, and interlocked with the train tablet or train staff instrument, so that if permission has been given for a train to approach, the SHUNTING STAFF cannot be obtained. The shunting staff (like the SHUNTING SIGNAL) only allows the driver to proceed into the section far enough to enable the rear vehicle of his train to clear the points. Under no conditions must the driver proceed ahead into the section without the correct STAFF or TABLET for that section. Where shunting staffs are provided, shunting is sometimes allowed under the rules for blocking back outside home signal. At places where either of these shunting facilities is allowed, starting signals are sometimes provided on the single line sufficiently far ahead of the points (as in the case of double lines), in which event it is common practice to place a home signal (for the opposite direction) just beyond the starting signal.

Locking is sometimes carried out between the staff or tablet instruments, and the starting signal levers, with a modification of lock and block working, so that the starting signal cannot be lowered until a staff or tablet has been obtained.

If it is desired to switch out a staff or tablet station, it is necessary to provide additional staffs or tablets, with instruments to suit, for the new long section to be used when the box is switched out. It is necessary that the two sets of instruments shall interlock each other, so that it is impossible for a tablet or staff for the long and short sections to be used by drivers simultaneously. It is of course possible to switch out more than one staff or tablet station if desired, and make the long section cover two or more short sections.

Boxes so switched out must have a device which permits of the signals for opposite directions being pulled to clear during the hours that the box is switched out.

Messrs. Sykes have a system employing lock and block working, in which staff and tablets are dispensed with. The instruments are designed with slides similar to the tablet instruments. The signalman at "B" pushes in his slide to release the slide at "A," the signalman at "A" on pulling his slide, after being released by "B," causes his starting signal to show CLEAR (an electric signal being used for this). The train, on passing over a treadle ahead of the starting signal, places the signal to danger, and it cannot be cleared again to allow a second train to enter the section until the first train has passed over a second treadle at "B," as in the usual lock and block method. The slide at "A," when pulled out, becomes back-locked until the slide at "B" has been restored to its normal position. The slide at "B," on being pushed in to release "A's" slide, becomes locked in that position until the train passes a treadle ahead of the home signal, when it can be restored to the normal position. For "B" to send a train to "A," the slides must be reversed in position, "A's" slide being pushed in and "B's" slide then being pulled out. Either signalman can push in the slide, but only one slide can be pulled out at any one time, and then only with the permission of the other signalman, whose slide must be in the correct "IN" position, and whose bell key must be pressed down, as in the case for an ordinary tablet instrument.

AUTOMATIC SIGNALLING

A system of signalling which has come into favour with increasing rapidity of late years is that of AUTOMATIC SIGNALLING, which may be considered as a further development of TRACK CIRCUIT BLOCK (see p. 63). With that system of block the state of the TRACK CIRCUIT, *i.e.*, "occupied" or "clear," was the sole criterion as to whether or not conditions were suitable for the STARTING signal to be pulled off. The system of "AUTOMATIC SIGNALLING" carries this arrangement a stage further and operates the signal directly by the track circuit without any intervening human agency; so that the signal will stand at "clear" all the time its controlling track circuit is clear and will indicate "stop" when its track circuit is occupied.

Automatic signals may be of the power-operated semaphore, or of the Colour light type, and may be used both for Double and (with certain additional safeguards) for Single lines. The great advantages of this system of signalling over ordinary BLOCK TELEGRAPH combined with manual signals are its freedom from the fallible human element, its simplicity, and its saving in operating labour costs.

A complete description of the ingenious devices incorporated in present-day AUTOMATIC SIGNALLING would, however, come under the head of Electrical, rather than Mechanical, Signal Engineering, and so would not be in place in the present volume.

REMINDERS FOR SIGNALMEN

TRAINS or engines which have been shunted from a siding to a running line, or from one running line to another, also trains or engines which have been sent forward to stand at the advanced starting signal, are liable to be forgotten by the signalmen unless some special means is adopted to remind the signalmen of the presence of the train. If the Block System is rigidly carried out the Block Instruments in such cases would indicate TRAIN ON LINE, the BLOCKING BACK SIGNAL would have been used in the case of shunted trains, and the TRAIN OUT OF SECTION Signal should *not have been given* in the case of a train which had arrived from the box in the rear, and been sent to stand at the starting signal. Some companies' rules, however, do not give this latter safeguard (see p. 58).

The Railway Clearing House Standard Rule No. 55 instructs one of the train men to go to the signal box as a reminder when the train has been kept at a signal on a running line for an unusual length of time.

Electrical and mechanical reminders are extensively used for the above purpose.

The simplest is a ring or plate which can be slipped over the Home Signal lever handle, and so prevent the catch being lifted to enable the signalman to pull the lever in order to send a second train forward, which might collide with the standing train. The signalman is instructed to place the ring on the lever of the protecting signal as soon as a train has been sent to stand on a running line.

Where the signal is some distance from the signal box a good deal of time would be lost by the train having to wait, after the line is cleared, until the train man rejoins it. Again, where there are many lines, it is objectionable to have several train men in the signal box as this is liable to hinder the signalmen in their work. It is also objectionable for the train men to have to cross several running lines on their way to the box, as the risk of personal injury so incurred is considerable, especially at busy boxes and during darkness.

To overcome these disadvantages of Rule 55 and to obviate the necessity for any of the train men going to the box as a reminder, an indicator—which may work in conjunction with a lock on the protecting signal, or a lock on the Block Instrument preventing a second train from being accepted—can be installed in the signal box, and can be operated either by track circuit or electric balanced bars (see Fig. 88).

Where the Lock and Block System is fully installed, no reminder is necessary, as full protection is afforded by the locking.

When balanced bars or track circuits are not desired a simple device fixed on the advanced starting signal post to be operated by the fireman is sometimes used. This takes the form of a press button or switch handle, with an indicator both in the signal box and at the signal, so that the fireman can see that the device works correctly. The apparatus should be arranged so that the button being pressed in, or the handle turned, indicates VEHICLES ON LINE, and locks the protecting signal (and the block instrument if required). The lowering of the advanced starting signal could be used to re-set the apparatus to the normal position, or the fireman could clear the indicator by turning the handle back as soon as the signal is lowered.

When a home or starting signal is more than 400 yards from the signal box it is now the practice to track circuit the rails in the rear of the signal. In the case of a home signal the track circuit operates an indicator in the signal box and locks the block instrument for the rear section. If the rear section is short it is usual to lock the starting signal of the box in the rear.

In the case of a starting signal the home signal in the rear is locked, and an indicator provided in the signal box.

Where these devices are installed an indicator is fixed on or near the signal for the purpose of letting the train men know that it is not necessary to proceed to the signal box to carry out Rule 55. This takes the form either of a "Diamond" sign of white enamelled iron with a black margin (as at A, Fig. 60), signifying that the presence of the train is automatically indicated in the signal box and that no action under Rule 55 is necessary; or of a so-called "D" sign (as at B, Fig. 60), signifying that one of the enginemen must communicate the presence of the train to the signalman by means of some device (usually a "Fireman's Press Button" or Telephone) fixed to the post carrying the "D" sign.

The foregoing description of the rules and methods of block working is by no means a complete record of all the rules necessary for the working of traffic under all conditions, such as emergencies and breakdowns, etc. Only the main principles of block working have been included for the purpose of showing how the Ministry of Transport Requirements with respect to the provision of an adequate SPACE INTERVAL are carried out in practice.

Paragraph 1 (Section B) and Section C of the Ministry of Transport Requirements refer to "Block Telegraph," and "Modes of Working Single Lines," and read as follows:

(SECTION B)
BLOCK TELEGRAPH

Paragraph 1. Apparatus to be installed for ensuring, by means of the Block Telegraph system, or by other approved method, e.g., automatic signalling, an adequate interval of space between following trains, and, in the case of junctions, between converging or crossing trains.

In the case of single lines, or sections of single lines, worked by one engine or motor vehicle (or two or more such engines or vehicles coupled together) carrying a staff, no such apparatus will be required.

On lines used purely for goods or mineral traffic, some other approved method of working may be substituted for Block Telegraph.

On passenger roads, exemption from block working in special conditions may be granted when essential for traffic purposes.

(SECTION C)
MODES OF WORKING SINGLE LINES

In the case of a single passenger line, an undertaking must be sent to the Ministry of Transport, through the inspecting officer, to the effect that one of the following modes of working single lines will be adopted, namely:

I.   *By train staff and train tickets in the mode described in the following rules, combined with the absolute block telegraph system.*

Rules for working the single lines between A, B, C, &c.

1.   *Either a train staff or a train ticket is to be carried with each engine or train to and fro, and for this purpose (one, two, or more) train staffs and sets of train tickets will be employed, e.g. :*

		Colour of Staff and Ticket.	Form of Staff and Ticket.
One between	A and B	Red	Square
" "	B and C	Blue	Round
&c.	&c.	&c.	&c.

2.   *No engine or train is to be permitted to leave or pass either of the staff stations A, B, C, unless the staff for the portion of the line over which it is to travel is then at the station; and no engine-driver is on any account to leave or pass a staff station without seeing such train staff.*

3.   *If no second engine or train is intended to follow, the staff is to be given to the engine-driver.*

4.   *If other engines or trains are intended to follow before the staff can be returned, a train ticket, stating "staff following," is to be given to the engine-driver of the first engine, and so on with any other except the last, the staff itself being sent with the last. After the staff has been sent away, no other engine or train in the same direction is to leave the staff station, under any circumstances whatever, until the return of the staff.*

5.   *The train tickets are to be kept in a box fastened by an inside spring, and the key to open the box must be the train staff, so that a ticket cannot be obtained without the train staff. The removal of the train staff must lock the box.*

6.   *The train staffs, the train tickets, and the ticket boxes are to be painted or printed in different colours, e.g., red for the line between A and B, blue for that between B and C, &c, the inside springs and the keys on the staffs being so arranged that the red staff cannot open the blue box, or the blue staff the red box, and so forth.*

7.   *The ticket boxes are to be kept in the signal boxes or in the booking offices at the staff stations.*

8.   *The sole person authorised to receive from an engine-driver, or exhibit or deliver to an engine-driver the staff or ticket is either the stationmaster, the inspector, the signalman, or the person in charge for the time at a staff station.*

9.   *In the event of an engine or train breaking down between two staff stations, the fireman or guard is to take the train staff, if with the train, to the staff station in the direction whence assistance may be expected, so that the staff may be at that station on the arrival of an engine. Should the engine or train that fails be in possession of a train ticket instead of the staff, assistance can only come from the station at which the train staff has been left. The fireman will accompany any assisting engines to the place where the engine, or train broke down.*

II.   *By divided train staff (without train ticket) combined with the absolute block telegraph system.*
In this case one-half of the staff will be marked "ticket" and the other "staff."
The rules for working a single line in this case will be similar in all respects to those above quoted, with the exception that the "ticket" portion of the staff is substituted when required for a train ticket. Each portion of the staff can be fitted with a key for controlling intermediate siding connections.
(N.B.—For light railway working, the block telephone in lieu of the block telegraph system will be accepted with either of the above-mentioned modes of working.)

III. With only one engine or motor vehicle, or two or more such engines or vehicles coupled together upon the single line or any section thereof at one and the same time.

Such engines or motor vehicles to carry the staff belonging to the line or section on which the train is travelling.

(N.B.—No ticket to be allowed under this mode of working.)

IV. By an electric token system, under which only one of the tokens applying to any section can be in use at the same time.

(N.B.—The approval of the Ministry of Transport to be obtained for the apparatus proposed to be used, and for the rules of working, which should be of a somewhat similar character to those detailed under mode of working, No. 1.)

V. By any other method approved by the Minister of Transport.

5
POINT CONNECTIONS

WHERE two lines join together, the rails have to be laid down so as to ensure that the vehicles can run on to either line as required. The junction is effected by means of "points." There are two rails, free to slide, which are planed down to a point. The position of these rails, which are usually called "switches" (sometimes also called "point tongues," or "switch blades"), determines the direction of a vehicle when it runs over them in the *facing* direction.

Points are generally classed as FACING POINTS or TRAILING POINTS. They are called "facing points" when a train travels towards them in the direction which admits of its being run *to* either of the *diverging* lines (see Fig. 112, A). A train running *from* one of the two lines in the *converging* direction is said to *trail* through the points; when the points are so laid that for trains running in the correct direction on the line in question, the points are trailed through, the points are then termed "Trailing Points."

Fig. 112, B, shows a passing loop on a single line of railway. The points lettered "a" and "b"

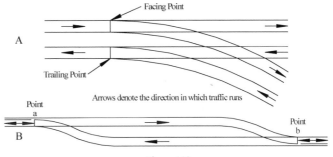

Figure 112
Facing and trailing points

act as both facing and trailing points. Trains running from left to right run over "a" in the facing direction, and over "b" in the trailing direction, but trains running from right to left run over "b" in the facing and "a" in the trailing direction. In cases of this description, where points may be run over in both directions, they are classed as "Facing Points."

For cross-over roads and the majority of siding connections, the points are laid down as trailing points, so that a train requires to move *backwards* to run over the points in their *facing* direction.

At any pair of points, should both of the switch blades be close against the main rails (these are called "stock" rails) (see Fig. 114, A), a vehicle in running over them in the facing direction would mount one or both of the blades, and possibly damage them; there is, besides, the probability of the vehicle coming off the rails. Should both the blades be standing *away* from the stock rail (Fig. 114, B), a vehicle running over them in the facing direction would drop off the rails into the 4-ft. way.†

Besides derailing the vehicle, the rails would most likely be damaged. This arrangement is sometimes used for TRAP POINTS where it is desired to derail a vehicle should it be moved too far, and where it is necessary to prevent the derailed vehicle from fouling adjacent lines.

---

† The distance between two running rails is 4 ft. $8\frac{1}{2}$ in., this being the standard gauge of all important lines in Great Britain, and is familiarly termed the "4 ft." The clear distance between the edges of the adjoining rails of the "up" and "down" roads of a double line railway is 6 ft., and is similarly termed the "6-ft." (Fig. 113.)

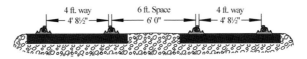

Figure 113
Four feet ways and six feet space

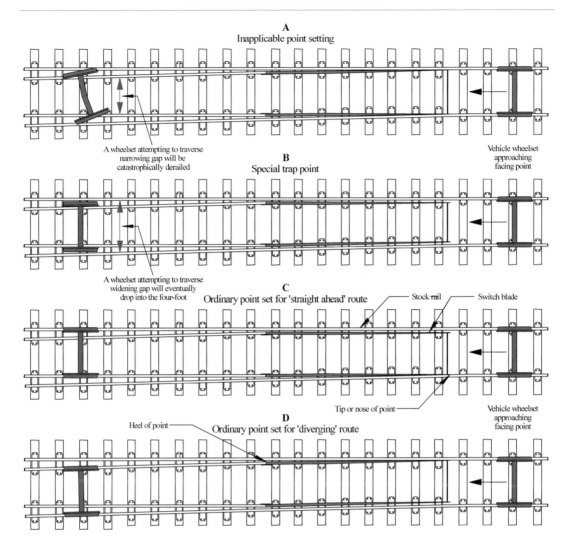

Figure 114
Setting of switch blades

Under ordinary circumstances, however, neither of these arrangements is required. At ordinary "points" *one* blade must be tight up against the stock rail (see Fig. 114, C and D), whilst the other blade must be standing away from its stock rail before a vehicle can run over them with safety. It will be seen that both blades must work together. To ensure that this shall be the case the Ministry of Transport Requirements state that there shall be not less than two connecting-rods between the blades, commonly called "stretcher rods" (see Fig. 115, A). These rods are designed so that the blade which stands open is about 4 in. from the stock rail at the tip of the point. At the first connecting-rod the blades can move about 3³/₄ in., and at the other end of the blade (called the "heel" of the points) the motion is, of course, *nil*. The blades are sometimes bolted at the heel to the full-sized rail by means of fish-plates, but with recent practice the blades are planed out of a full length rail, in which event there will be a few keyed chairs on the full section end of the blade before the fish-plate is reached. These are termed "heel-less" points, and any motion of the blades at the tip or "nose" *slightly* bends the rail.

With trailing points the connection to the lever working the points is made up at the first stretcher rod (counting from nose of points), where the travel is about 3³/₄ in. In Great Britain the connections working the points are almost invariably rods, although on the Continent wires are very extensively used.

The rods, generally called "point rods," are of various sections (Fig. 116), but perhaps the most ordinary type is round tube "rodding" about 1¹/₄ in. external diameter, and about ⁷/₈ in. internal diameter. Several companies are now adopting round solid point rodding, whilst channel section rodding 1⁵/₈ in. by 1¹/₈ in. by ¹/₄ in. is very much favoured by some companies.

Point rodding is supplied in lengths of about 15 ft. to 18 ft., and to form a run of rodding the lengths have to be joined together. There are several types of joints. For hollow rodding a solid plug

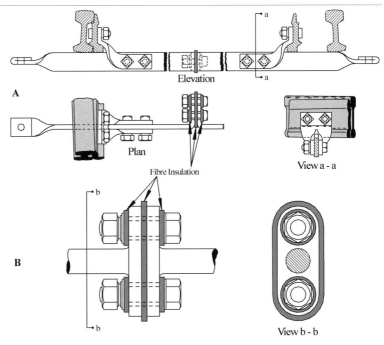

Figure 115
Insulated stretcher rods

is welded into the tube at each end; this plug is screwed with the standard Whitworth thread (seven threads per inch), and collars or couplings, with internal threads to suit, complete the joint (Fig. 117, A). The couplings are tightened up with pipe tongs, and the threads are so screwed on the rod that it requires some force to tighten up the last three turns. This gives sufficient grip, and prevents the couplings unscrewing in service.

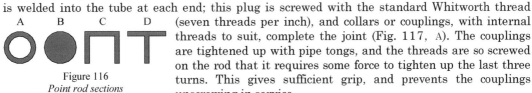

Figure 116
Point rod sections

Another method of making a joint is to rivet a socket into one end of the tube, this socket having a slot cut in it to receive a cotter. The other end of the tube is also slotted, and when the spike of the socket is inserted in the end of the tube to be joined, the slotted holes allow a cotter to be driven through the tube and socket (Fig. 117, B). The end of the cotter is split to prevent its accidental withdrawal, and a small projection is formed on the top of the cotter to prevent its falling through the slot should it wear very slack through much use. With solid rods sometimes the screwed joint is used, or the ends are formed into tags as in Fig. 117, C, to allow of their being bolted together. With channel or T section rodding the ends are generally butted together and bolted up with fish-plates (Fig.

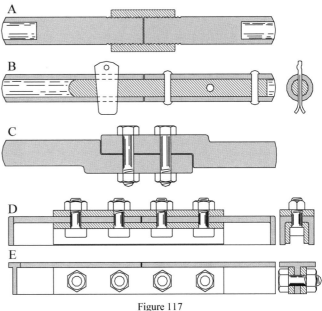

Figure 117
Point rod couplings

117, D and E).

With modern signalling the Track Circuit (see p. 47) is becoming more and more universal, so that the fixing of insulation is a very appreciable item in mechanical signalling.

The insulated rail joints (which separate electrically the track circuited rails from the remainder of the line) are usually installed by Permanent Way forces, but it is also necessary to insulate all signal connections to and across the track circuited rails.

For the insulation of

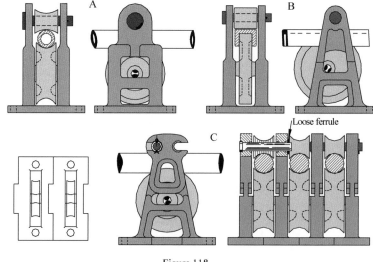

Figure 118
Point rod rollers and frames

rods operating points, bars, detectors, etc., a joint such as that shown in Fig. 115, B is used. This is composed of two wrought iron face plates, each welded to the point rod, and separated from each other by vulcanised fibre packing, the connecting bolts being insulated from the face plates by fibre collars and washers. For the insulation of facing point and other stretcher bars the bent-up ends of the bar itself serve as face plates, the arrangement being otherwise similar to rod insulation (see Fig. 115, A).

Fig. 139 shows the positions where insulation is necessary in a facing point layout, the underlying principle of the arrangement being that there must be insulation in any metallic connection from one track circuit rail to the other, and also in any metallic connection between either rail and "earth" (a run of point rods or a mechanical detector is considered as "earth"). Where signal wires cross under insulated rails a small piece of board should be nailed over them from sleeper to sleeper, or other means taken to ensure that when the wires are pulled they do not rise up and make contact with the bottom of the rail.

The rods are supported on rollers; these are of various designs to suit the various types of rods used. Whatever the type of roller, however, it is now universal to arrange the bearings of the roller wheel on the anti-friction principle. This is effected by allowing the pin forming the axis of the wheel to roll on its bearing, the bearing hole in the supporting frame being elongated for this purpose. The roller frames are designed so that a roller can be removed if damaged in service (Fig. 118), and a new one substituted without taking up the frame. The frames are also designed to allow of extensions being effected easily. Generally the framing for one roller is in halves, each being a separate casting made to check into each other, and held down to timbers with spikes or coach screws. If there is one rod laid down, each roller requires two frame castings. To lay an additional rod only one additional frame casting is required, as the roller fits into one of the existing frames on one side and into the additional frame on the other side. With some types (Fig. 118, A and B) a long pin runs through a hole in the top of all the frames in one row, and by this arrangement the frames are kept in line; but in the event of additional rollers being laid down, the existing pins have to be replaced by longer ones. With other types the connecting pins only couple up two frames (Fig. 118, c), the first pin being on the rear of the frame, the next pin coupling up the last frame and another one, and so on, the pins being placed first on one side of the frame and then on the other. These pins also act as bearings for small top rollers. With this arrangement all that requires to be stocked in the shape of standard castings for any width of rodding is *one* pattern roller frame, one pattern roller to support the rodding, and one top roller to assist in guiding the rodding, together with one standard pin for binding the frames together.

A common pitch for the rollers is 2⅝in. centre to centre. Stools for supporting the rollers (Fig. 119) are made of creosoted wood or concrete, the sizes for the top of the stools being about, for 1 to 3 rods, 17 in., for 4 to 6 rods, 2 ft. 0 ½ in.

The stools are pitched about 7 ft. 6 in. or 8 ft. apart for straight runs of rodding, but if a sharp curve has to be taken without the aid of relief cranks, then the pitch has to be reduced to meet the case, and is usually 5 ft. Rodding should, however, be kept as straight as possible to ensure good working.

To connect a run of rodding to the stretcher rod of the points, a special end termed a JOINT is used. The joint is, as a rule, welded to a length of solid rod, and the other end of the rod either screwed or otherwise finished off to suit the type of point rod coupling adopted (Fig. 120). These joints are of various sizes to suit the particular apparatus requiring to be connected up. Usually for point rods the pins for the joints are from ⅞ in. to 1 in. diameter. It is convenient to have some of the joints adjustable for length, and this is effected by having a

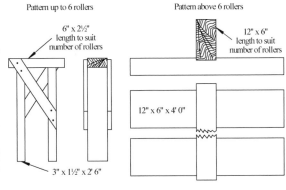

Figure 119
Timbers for roller frames

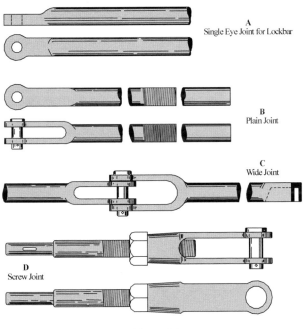

A
Single Eye Joint for Lockbar

B
Plain Joint

C
Wide Joint

D
Screw Joint

Figure 120
Rod joints

screwed joint, the end of the joint being tapped with a thread to fit on to a screwed piece of rod (Fig. 120, D).

The joint which connects to the stretcher rod is generally of the plain joint type, not screwed. The end of the rod to which the joint is welded, however, is screwed with a left-hand thread, and screws into a socket. The opposite end of this socket has a right-hand thread with a piece of rod screwed to suit. The socket or "barrel" as it is sometimes termed, has a nut formed on it or some other convenient device to enable it to be rotated.

The effect of turning the barrel is to lengthen or shorten the rod (Fig. 121). This device is called an "adjusting screw" or "union screw coupling." There are several designs for this article. The type to be preferred is one which *exposes* the portion of the thread which is not actually enclosed in the tapped portion of the barrel. An enclosed type does not permit of inspection, and there is danger of the threads becoming rusted and liable to fail. Where the end of the thread is exposed it enables the condition of the thread to be detected before it can become unfit for service. All adjusting screws and screw joints are fitted with check or lock nuts to prevent their working loose.

For transmitting motion round an angle of more than a few degrees, cranks are used, the most common type being right-angle or L cranks (Fig. 122). The cranks are made of wrought iron, cast iron, or cast steel. Forgings are very convenient in practice, as it is possible to set the arms up or down as may be required to suit a particular case. Cast-iron and cast-steel cranks are usually made with high and low arms, as it is, of course, impracticable to set these on the

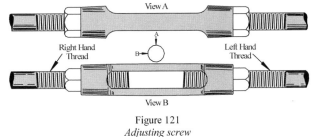

Figure 121
Adjusting screw

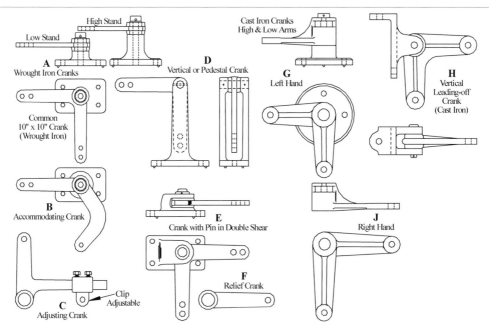

Figure 122
Cranks and stands

ground. The boss or bearing of the crank should be of ample size to ensure a long term of use before it becomes too slack. The arms are made of various lengths, but for ordinary point work 10 in. from centre of crank to centre of pin is most common. With forgings it is usual to have two holes in each arm, one at 10 in. and another at 8½ in. (see Fig. 122, A, D and E). This allows the travel of the rodding to be adjusted by moving the pin from the 10 in. hole to the 8½ in., or *vice versa*, as may be required. With cast-iron cranks, however, it is difficult to design a strong crank with two holes in each arm, and there are, therefore, several pattern cranks in use with arms 10 in. by 10 in., 10 in. by 9 in., and 10 in. by 11 in. (Fig. 122, G, H and J). If the travel of the rodding requires altering at a particular crank it is necessary to put down a different crank to suit the case.

The stands for the cranks are, as a rule, of cast iron, the centre pin being cast solid with the stand. In some designs the centre pin is chilled, while others have a turned wrought-iron centre pin. Most designs have the centre pin so that the crank can be lifted straight off, the pin being in single shear (see Fig. 122, A and G); but some companies have a stand with the pin in double shear (see Fig. 122, E). This makes a stronger bearing, but not quite so convenient in use. It is sometimes necessary to use cranks in a vertical position, with upright stands to suit (Fig. 122, D and H). These stands are made of cast iron with two upright sides, with a turned wrought-iron pin, which pin is in double shear. Vertical cranks are generally used for leading a run of rods out of the signal box from the levers working them, and are either fixed with coach-screws to timbers fastened to the bottom floor of the signal box, or to timbers built into the face-wall of the signal box.

Ordinary horizontal cranks are generally fastened to timbers, which have been creosoted, the size being 12 in. by 6 in., with the length to suit the case. It is often possible to fix several cranks on one length of timber. Coach screws about 7/8 in. diameter are commonly used for holding down. The top timbers are spiked down to cross-pieces which are firmly bedded in the ballast. These stands are similar to those used for wheels (see Fig. 99). It is most important that the crank timbers shall have a firm hold in the ground, as any movement of the timbers puts the point connections out of adjustment.

FACING POINTS

WHERE points are run over in the facing direction by passenger trains, the Ministry of Transport Requirements call for the points to be fitted with facing point locks, locking bars and detectors.

The facing point lock prevents the points from being moved either by accident, or by anyone attempting to operate them from the signal box, so long as the lock bolt is in the bolted position; also it ensures that the point blades are held tightly against the stock rail, even though the rodding working the points be disconnected or damaged. The British Standard lock is made up of four main parts.

1. *The Lock Stretcher Bar.*—This is a piece of iron bolted to both of the switch blades about 5 in. from the nose,† with one or two holes or notches in it to allow the bolt to enter when the points are locked (see Figs. 115 and 123). The standard stretcher is of a single piece of metal of 3 in. by $^5/_8$ in. section, but in some designs there are two pieces, one fastened to each blade (Fig. 124, D and F), with both ends coming together in the middle of the 4 ft., so that the bolt enters through holes in both pieces. These are termed "split stretchers."

2. *The Lock Bolt.*—This is commonly a simple bar of iron about 2 in. by 1 in. section (Fig. 123). One end has a hole in it to allow it to be attached by means of a joint to the rodding working it. This bolt is moved in through holes or notches in the lock stretcher when the points are bolted.

3. *The Lock Shoe.*—This is a casting designed to accommodate both the bolt and the lock stretcher, allowing each to move only in its proper plane. The lock shoe is fastened down with coach screws or through bolts to the same sleeper as the first slide chair of the points. The lock shoe has a cast lid which is held in place by the pin C (Fig. 123); the pin E, serves to keep the lock bolt in position.

4. *The Detector Slide.*—This is a casting which slides in a recess in the lock shoe and is designed with projections B (Fig. 123) which engage with the lock bolt so that when the latter has been moved into the notch in the stretcher bar, the bevel faces A drive the detector slide sideways and so operate the detector (see Fig. 138).

It is essential that the lock shoe shall be firmly fixed to the sleeper, as otherwise, although the bolt is in through the stretcher, the points cannot be held tightly against the stock rail; it is also very necessary that the first slide chair to which the stock rail at the nose of the points is bolted shall be firmly fastened to the sleeper; otherwise, although the point blades be held in the correct position, the stock rail might move away from the blades under the pressure of the wheels. In order to keep the stock rails to gauge at the points the Ministry of Transport Requirements call for Stock Rail Gauge Ties to be fitted. A type of tie which used to be fitted had no connection with the facing point lock or the sleeper, so that while it held the stock rails firmly together it did not prevent *both* of the stock rails from moving together and so leaving a space between the blades and the stock rail. It is now, therefore, considered desirable to fit all points, and especially facing points, with a steel tie plate which lies on top of the sleeper supporting the first slide chair (Fig. 125). To this tie plate strips are riveted which are fitted to the slide chairs outside, so that they form ties and prevent these chairs from spreading; and as the stock rails are bolted to the slide chairs, they also are prevented from spreading. The lock shoe is fixed on the top of this tie plate, the holes for the coach screws or holding-down bolts being drilled in the plate to suit, as also for the spikes or bolts of the slide chairs.

In fitting the bolt to a set of facing points it is essential that the bolt shall only enter the hole in the stretcher when the blade is held *tightly* against the stock rail, and on no account must the end of the bolt be rounded or tapered off to allow it to enter easily and force the blades into position, as should this be done it is quite possible for some foreign matter to become jammed between the blades and the stock rail; and should the bolt be able to force the blades against the stock rail the tip of the points would be forced open. Should the point rod be unable to place the blades in the correct position it must be impossible for the bolt to be moved through the hole in the lock stretcher.

In this way it will be seen that the bolt acts as a detector to the points, as, should the points not be properly "home," the bolt cannot be inserted. For easy working $^1/_8$ in. clearance between the stretcher and the *top* and *bottom* of the bolt can be allowed. The idea of using a split stretcher, as mentioned previously, was to detect both blades, thus ensuring that not only was one blade firmly

---

† This dimension depends on the type of switch blades employed.

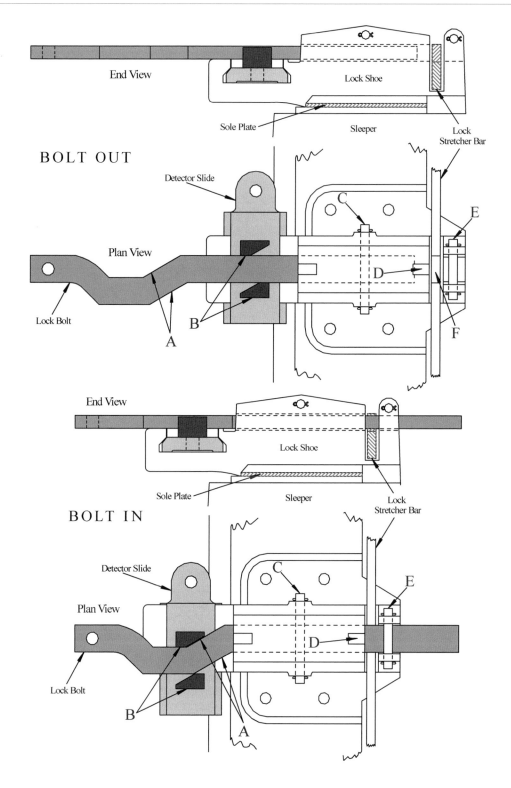

End View

Lock Shoe

Sole Plate

Sleeper

Lock Stretcher Bar

BOLT OUT

Detector Slide

Plan View

Lock Bolt

A

B

C

D

E

F

End View

Lock Shoe

Sole Plate

Sleeper

Lock Stretcher Bar

BOLT IN

Detector Slide

Plan View

Lock Bolt

C

E

D

B

A

Figure 123
British Standard facing point lock

held against the stock rail, but that the other corresponding blade was moved away from its stock rail to give clearance for the flange of the wheel to pass. With modern facing point gearing, however, the bolt is not regarded as being an efficient detector, as it is possible for the rodding working the bolt to be forced, although the bolt might not have entered into the hole in the stretcher, hence the split stretcher is falling into disuse.

One of the main features of the B.S. Facing Point Lock is the ease with which it allows the essential clearances to be inspected, and the lock cleaned and oiled. The clearance of the bolt from the sides of the notches in the stretcher bar can be examined (at F, Fig. 123) and the position of the bolt when withdrawn can be checked by the recess D, both without removing any pins. By removing pin C the cover can be taken off without

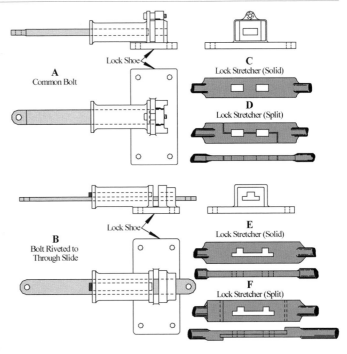

Figure 124
Facing point locks and stretchers

affecting the security of the points if they are bolted, as the pin E will keep the bolt from rising out of its place. When the signalman withdraws the bolt it can, without disconnecting it from its driving rod, be raised out of the shoe for the latter to be wiped out and oiled.

With some of the "Economical" or escapement types of facing point layouts, a lock of a different type (see Fig. 124, B) is required, in which the bolt moves in one direction only in order to Unlock and Relock the points (see p. 89). A pattern of bolt lock which may be used for Facing Points controlled by ground frames on single lines, or for mechanical ground frame controls or other general purposes is illustrated in Fig. 124, A.

Facing points are in most cases equipped so that the bolt can enter when the points are either NORMAL or REVERSED, and the lock is then termed a BOTHWAYS lock. This is the case of an ordinary running junction, where the trains can be sent in two directions. There are, however, many cases where the train is only required to run in one of the diverging directions, such as the points at a passing place on a single line (see Fig. 1 1 2 , B). At a place as shown a train will only require to run in the directions indicated by the arrows. The points A and B are both FACING POINTS in one direction, but TRAILING POINTS in the reverse direction. A train must not be allowed to run from the single line on to the double portion in the *wrong* direction, hence the facing point locks at A and B lock the points ONE WAY only. That is, the bolt can only enter when the points are in their normal position; if the points are in the reverse position to allow a train to run *from* the double line portion on to the single line, it is not possible to insert the bolt.

Lock Bars.—The lock bar is a device for preventing the signalman from unbolting the facing points whilst a train is standing on or running over them. There are two general classes of lock bars, inside and outside. The inside lock bar is fitted to the rail so that it engages with the *flange*

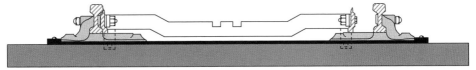

Figure 125
Steel tie plate and B. S. facing point stretcher

of the wheel (Fig. 126), whilst the outside lock bar engages with the *tread* of the wheel (Fig. 127).

A lock bar generally is made of a length of T section iron, the usual size being 2 in. by $1^1/_2$ in. by $^5/_{16}$ in.

The length of the bar depends on the length of the greatest distance between the wheels of the vehicles in use on the line. This, of course, varies greatly on different railways, the longest bars used being about 40 ft. and the shortest about 18 ft. The

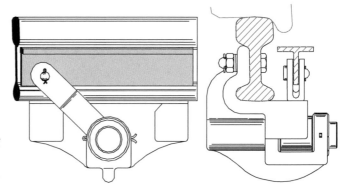

Figure 126
Inside lock bar with cast iron clip and stops

British Standard is 40 ft. long. The bar is supported by short levers which work on "clips" or "hangers" fitted to the rail, so arranged that when the bar is moved in a direction the same as its length it rises slightly. This is effected by the supporting levers moving in an arc of a circle as shown in Fig. 128.

Normally the bar is down clear of the flanges of the wheels, but when it is moved it rises up higher than the flanges, so that should any wheels be over the bar it is impossible for the bar to be moved up. The depth of a wheel flange is about $1^1/_8$ in., and, allowing for wear, both in the tread of the wheel (making the flange deeper) and on the tread of the rail (letting the flange come lower down than when new), it is usual to arrange for the normal position of the bar being about $1^3/_4$ in. below the top of the rail, when both the bar and rail are new. This allows a clearance of $^5/_8$ in. between a new bar and wheel flange when a new rail is in use, so that the bar moves up this $^5/_8$ in. before it comes in contact with the flange of the wheel.

When the bar is fitted on the outside of the rail the top of the bar is very slightly below the top of the rail when both are new. When the bar is moved it rises above the top of the rail, the rise of this bar being about the same as in the case of the inside bar. The outside bar is not satisfactory unless the width of the tyres is considerably greater than the width of the rail; as it is that portion of the tread of the wheel overhanging the rail which prevents the lock bar from being moved. With the section of rail adopted as the British Standard Section it is not desirable to use outside lock bars. To ensure that the bar shall always come in contact with the wheel, the bar is inclined slightly, so that it moves in towards the rail when it rises (see Fig. 127). Unless the fittings are maintained in first-class condition, and unless the rails are kept well to gauge, there is some risk of the lock bar being forced over, *i.e.*, it may move away from the rail as it rises.

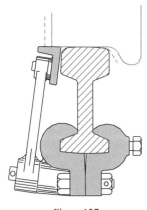

Figure 127
Outside lock bar

Angle section bars are often used for outside bars, or a special section bar may be adopted.

The levers on which the bar works are generally made of wrought iron, and are about 6 in. from centre of bearing to centre of pin. The "hanger," or "clip," which is bolted to the rail, and which supports the levers, is commonly made of wrought or cast iron. It can be designed either for the bearing pin to be in single shear or double shear, the latter being preferable. It is sometimes designed

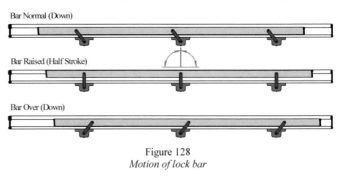

Bar Normal (Down)

Bar Raised (Half Stroke)

Bar Over (Down)

Figure 128
Motion of lock bar

with stops to limit the movement of the levers and sometimes is fitted with a guide to prevent the bar from being forced away from the rail. The British Standard lock bar is, however, provided with separate castings bolted to the rail which act both as stops and guides, so that these are not required on the clips. The clips are usually placed between each sleeper, and unless the section of the bar is extremely stiff the greatest distance between the hangers should not exceed 4 ft. Within reasonable limits a bar works easier for every additional hanger.

It sometimes occurs that the lay-out of a station or junction does not permit of sufficient space being provided in front of the facing points to accommodate the lock bar. In this case either two outside bars must be fitted alongside the points, one on each stock rail, or inside bars must be fitted to the point blades themselves. Outside bars offer no difficulty provided the section of rail is suitable for them, but, as previously mentioned, they are not nearly so efficient as inside bars.

With inside bars fitted on the point blades some arrangement

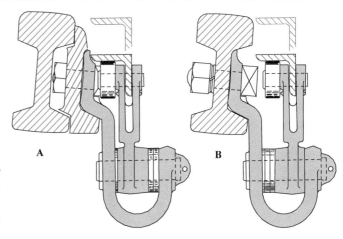

Figure 129
British Standard clip with angle iron bar

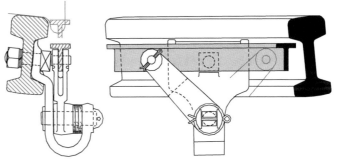

Figure 130
British Standard clip with tee iron bar

must be made for supporting the bar on the blades. If the lock bar is 40 ft. long, and the point blades 20 ft. long, practically one-half of the bar is fitted to solid keyed rails and the other half is fitted to the moving blade. Where the blade is the full section of the rail the ordinary pattern clips

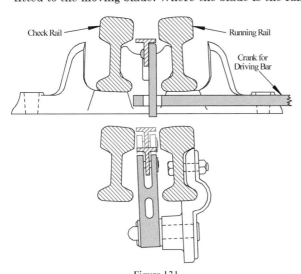

Figure 131
Lock bar between check rail and running rail

can be used, but where the planed portion of the blade is encountered, special clips have to be adopted to ensure that the bar is maintained at the correct distance from the running edge of the rail. Fig. 129 shows the British Standard Switch Clip, which is so designed, with steel washers and an unsymmetrical lever, that by varying the positions of the washers and turning the lever round in the clip, the bar can be suited to the end of the point blade (as at Fig. 129, A), or to any intermediate position back to the full section of rail (as at Fig. 129, B). Further, by substituting a plain pin for the roller and pin at the top of the lever, this clip is suitable (after re-arrangement of the washers) for fitting a T section bar to ordinary rail (see Fig. 130).

This figure also illustrates the relative position of a new wheel flange and the bar when in the normal and raised positions.

Where the lock bar requires to be fitted between the running rail and a check rail it is necessary to adopt a narrow section bar not more than 1 ¹/₂ in. wide.

The bars are driven by studs bolted to the web of the bar, to which an eye joint is fitted (see Fig. 120, A). With the bar between the running rail and the check rail this stud cannot be used, and a long stud is substituted, extending downwards below the level of the rail, which engages with a crank in a slotted hole (Fig. 131).

MECHANICAL DETECTION

THE purpose of providing a detector is to ensure that the points have been placed in the correct position for the train before the signal applicable to that pair of points can be pulled to "clear."

With facing points the bolt ensures that the points are tight home, but in most cases it does not ensure that the points are set in the correct direction. Apart from this it is quite possible for the connections between the signal box and the bolt to fail, in which case the bolt cannot be relied upon for efficient detection. Mechanical detectors are commonly divided into two classes—rod detectors and wire detectors. In the first mentioned, which are now not much used, the connection from the signal box is composed of point rodding up to the facing points;

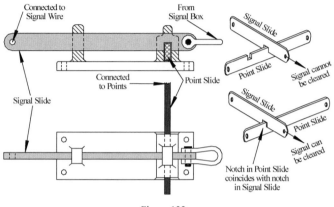

Figure 132
Slide detector

while with the wire detectors ordinary signal wire is used exclusively. If one signal is required to detect several points, electrical detectors are to be preferred.

Whichever class of detector is used, it is necessary to connect each point blade to the detector, and the Ministry of Transport now require the facing point bolt to be detected. In the detector there can be either (1) one slide for the points and one for the bolt, in which case the rods from the two blades have to be combined through the medium of a "combining" bar, or (2) preferably there are three distinct slides in the detector, one for each point blade, and a third for the bolt. The point and bolt slides have notches cut therein to register the position of the points (Fig. 132). Through this notch moves another slide worked by the connection to the signal. It is arranged so that unless the points are in the correct position for the signal to be cleared, and are correctly bolted, the notches in the point and bolt slides do not coincide with the notch in the signal slide, and this prevents the signal slide from being moved to pull off the signal. The signal slide can be of various designs. For rod-operated slides it often takes the form of a bolt (Fig. 133). For wire-operated slides it may be a plain slide working in guides with a hole at each end for the wire

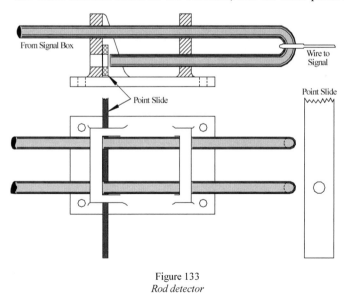

Figure 133
Rod detector

connection, as in Fig. 132, or it may be designed to move in an arc of a circle with an arm to which the signal wire is connected, as in Fig. 134. Allowance should be made for at least 6 in. travel of the signal wire at the detector. Whatever type of wire detector is employed, it is desirable to provide some form of counterbalance if the detector is to be fixed many yards from the signal post. If no means for counterbalancing the detector is adopted the temperature variations in the length of the wire between the signal and the detector will be liable to cause failures, but in the event of this connection breaking when the signal is "off " or of the signal sticking in the "off " position, the slide will have no weight to pull it back and will stick in the pulled position, thus holding the points and by preventing their being reversed until the failure is attended to, will

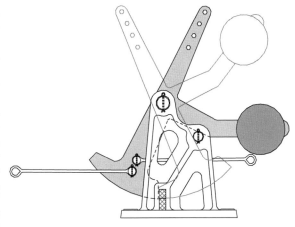

Figure 134
Counterbalanced wire detector

provide a valuable additional safeguard. With rod detectors, should the rodding break, unless some device is adopted to assist in moving the rodding to the normal position, it is likely to cause the signal to hang "off " instead of flying to danger as required by the Ministry of Transport Requirements.

Detectors are now commonly fixed to all points over which train movements are controlled by signals; hence many trailing points are fitted with detectors where trains back over them. They are adopted at all cross-over roads and siding connections to main lines. In these cases rod detectors are not used, as the signal usually detects a pair of points which are close to it. With disc and ground semaphore signals the detector is often incorporated with the counterbalance lever of the signal; if not, it is usual to use a short length of rod to connect the signal slide of the detector to the counterbalance lever of the signal; thus the one counterbalance weight serves to put the signal to danger and replace the signal slide of the detector.

ELECTRIC DETECTION

IN a complicated station yard it often happens that one signal is required to detect several sets of facing points, and also conversely that a particular set of facing points is detected by several signals.

It is frequently very difficult to fix mechanical detectors in satisfactory positions, and after they have been fixed they are often a source of trouble owing to the difficulty in correctly adjusting the wire between the different detectors. To assist the working of the signals, electric detection is now being employed at complicated places by many companies.

Electric contacts are fitted to the facing points, one set of contacts is closed when the points are normal whilst another set of contacts is closed when the points are reversed, the normal contacts being opened by the motion of the points.

With one system of electrical detection each signal has its own circuit outside, and unless all the points are in the required position for the train to run in obedience to that particular signal, an electric lock on the signal lever prevents its being pulled (see Fig. 135). With this arrangement a large number of electric contacts for the signal lock circuit are scattered about the

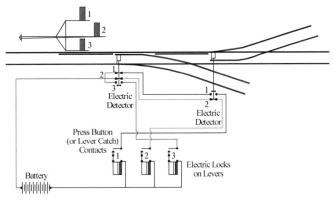

Figure 135
Schematic diagram for electric detection

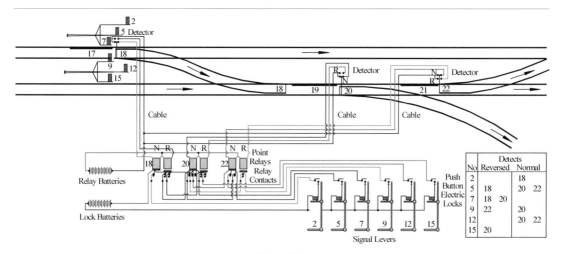

Figure 136
Schematic diagram for electric detection using relays

station yard, and in the event of a failure some difficulty is occasionally experienced in tracing out the signal circuits.

To simplify the circuits and concentrate as many contacts as possible in the signal box, electric relays are almost invariably used. The relays, if "polarised," have three positions for the armature (or two independent non-polarised relays can be used) to correspond with the position of the points "normal," "half-over" and "over." The armature of the relay has contacts on it to correspond with the signal circuits of the points requiring to be detected (see Fig. 136).

The objection might be made to the use of relays that, should an armature fail to act, it would be possible to pull the signal off with the points lying in the wrong direction. This, of course, could only happen if two failures occurred at once, *viz.*, the failure of the armature and a failure of the point connections, and with modern relays the contingency is so exceedingly remote as to be negligible.

With electric detection great care must be taken to see that the contacts are correctly adjusted. It is essential that the points shall not only be in the correct position, but that they shall be tightly held in that position; hence the contacts must be adjusted so that if the point blade does not lie tightly against the stock rail the contact will be open. If, however, the adjustment is made too fine, there is considerable risk of the blades moving very slightly during the passage of the train and breaking the contact. If the bolt of the facing points is also d e t e c t e d , the detection of the blades need not be so fine, as the fact of the bolt being in ensures that the point blades are tight against the stock rail.

It should be noted that if the detector is simply used to lock the signal lever, it is of no moment should the contact be broken whilst the train is running over the points, as the signal lever will then be pulled; if, however, the detector contact c a r r i e s current to maintain the signal i n t h e clear p o s i t i o n , as would be the case if a reverser or electric signal c u r r e n t flowed through it, it is very important that the signal shall not be put to danger owing to the detection being too fine.

CONNECTING UP FACING POINT GEAR

THERE are many arrangements for connecting up the bolt, bar, detector slides, and point blades. The usual method is to have one lever to work the points, as in the case of trailing points, the points, of course, driving the point detector slide; then a separate lever is used to operate the lock bar and bolt. Formerly the bolt sometimes was connected by means of a T crank to the lock bar, and a connection from the signal box was taken to the arm of the T crank (see Fig. 137), so that both bar and bolt worked simultaneously. Thus should there be a train on the bar it would be impossible for the bolt to be withdrawn. Should, however, the connection between the T lever and the bar fail, it would be possible for the bolt to be withdrawn without the bar moving. To prevent

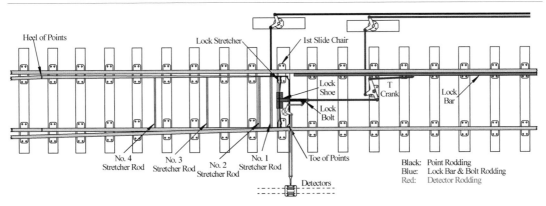

Figure 137
Facing point layout with T crank

this it is now usual to connect the bolt to the bar by means of a straight lever, and to make a separate connection by means of an L crank to the bar, so that the motion of the rodding from the signal box is transmitted to the bolt *via* the bar (Figs. 138, 139), hence should the bar not move it is impossible for the bolt to move. This arrangement is theoretically better than the one using the T crank, but it employs more joints, and is therefore more costly to install and maintain.

For all new work and alterations it is now necessary to detect the bolt in addition to the point blades. This is done by running a rod with an additional detector slide in the detector, so that unless both the blades and the bolt are all correctly placed, the notches in the slides do not coincide to allow the signal slide to move (Fig. 139).

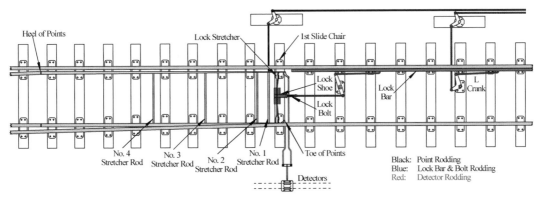

Figure 138
Facing point layout with L crank

Where a bolt detector is used it is essential that the connection shall be taken off the bolt itself, and not off any of the driving pins or connecting rods. The new British Standard Facing Point lock provides for the connection to the bolt detector being operated by a bevel cam forming part of the bolt itself.

With the lock bar and the points working on separate levers it is quite possible for the points to be moved under a running train *if the bolt has not been put in before the train arrives on the bar*.

It is arranged, by interlocking the levers in the signal box, that the signals cannot be pulled to allow a train to run on to the points, unless the bolt lever has first been pulled; but should the train over-run the signal, or should the signal be put back to the danger position after the train has passed it and before the train has reached the bar, it is possible for the train to be split on the facing points.

The latter contingency cannot happen if the signal is close to the facing points, as there would

(Continued on page 89)

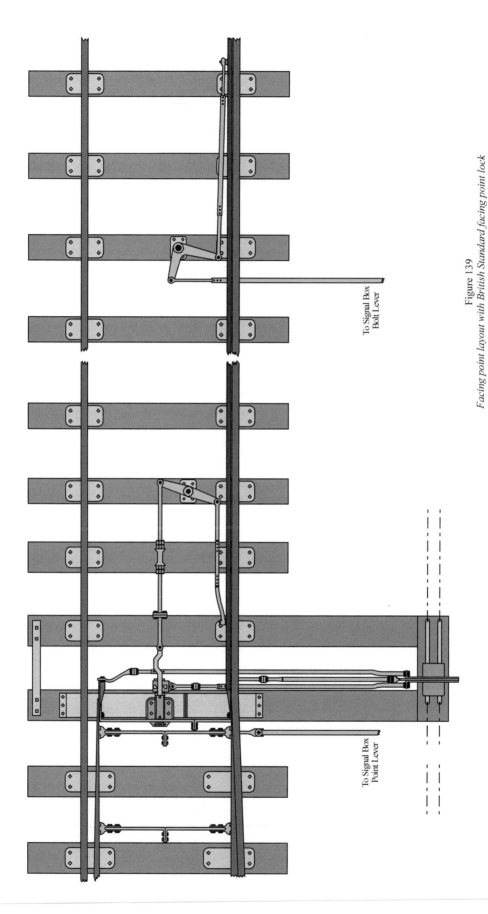

To Signal Box
Bolt Lever

To Signal Box
Point Lever

Figure 139
Facing point layout with British Standard facing point lock

(Continued from page 87)

not be sufficient distance for the train to pass the signal without being on the lock bar. Hence, unless some special arrangement is adopted to hold the facing points, the signal giving permission for the train to run over the points should be close to them. In the event of its not being so, an intermediate bar is fixed and interlocked with the signal and the facing point lock bar, or electrical locking devices may be operated by the train, which maintain a lock on the facing point bar until the train has cleared the points.

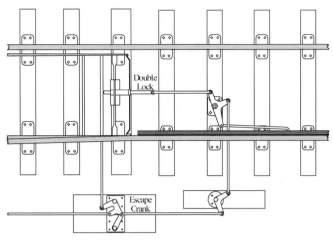

Figure 140
Simultaneous working of points and lock bar by escape crank

There are several arrangements in use for working the points and bar simultaneously by means of one connection from the signal box. These arrangements are, as a rule, cheaper than the separately operated points and bar gears, but are harder to work. They have, however, the advantage that it is impossible for the signalman to move the points whilst there is a train over them, as in all cases the bar moves every time the points are worked.

With these devices the points are always bolted except when they are actually moving. The first part of the travel of the connection from the signal box moves the bar and pulls out the bolt (or its equivalent).

When the bolt is out the points are moved across, either by means of an escape crank, as in Fig. 140, or some cam motion, as in Fig. 141, and then the bolt is returned to the bolted position, the bar being also lowered to its normal position.

In some cases the bolt moves out, and then in again, whilst in other types the bolt moves in one direction only, having two projections on it which engage with the stretcher to hold the points; one projection holding when the points are in one position and the other projection when the points are reversed (see Fig. 124, B). These arrangements are most commonly used where power working is employed to move the connections, but with ordinary manual working it is unusual to adopt combined point-and-bar gearing, unless economy of first cost is essential. It might be mentioned, however, that one large English railway company has a combined point-and-bar apparatus as its standard facing point lay-out.

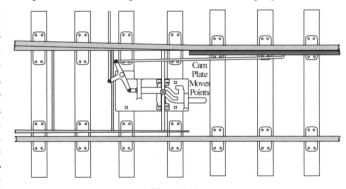

Figure 141
Simultaneous working of points and lock bar by cam plate

When the bars have to be placed alongside the stock rails or point blades instead of in front of them, as usual, two bars are employed, one bar protecting the points when the train runs in one direction and the other bar serving when the points are reversed.

Only one bar, therefore, is used at a time, so that it is possible to employ a "disengager" or "selector" which allows the bar not acting as a protection to lie idle, the "disengager" being operated by the points (see Fig. 142). Another device consists of two separate connections from the signal box, one for each bar, and a double bolt in the 4 ft.; one bolt working with each bar, as in Fig. 143, the interlocking being so arranged that each signal is released by the bar which protects

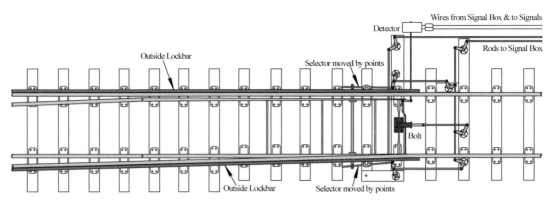

Figure 142
Outside lockbars with selector (disengager)

its particular route.

When power is employed to work the bars and points, it is common to connect both bars to the same motive power. With ordinary manual power, in many cases two bars are too heavy for operating, and two levers are then required.

With 50 ft. bars of stiff section, one bar alone is sufficient to move when it is any distance from the signal box. Several devices have been adopted to ease the motion, some taking the form of cranks with varying leverages, others employing some counterbalancing device, either a simple counterbalance weight or spring balance gear.

Either of these devices considerably eases the working of the bar.

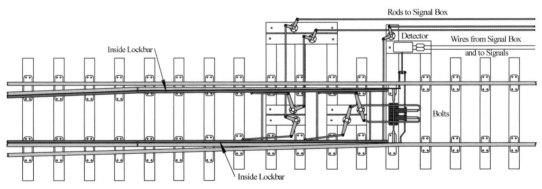

Figure 143
Inside lockbars with double bolt

CONNECTIONS FOR MOVABLE DIAMOND CROSSINGS

THE M.O.T. Requirements limit the angle of a fixed diamond crossing to a maximum flatness of 1 in 8. This seriously limits the speed of trains running round junctions branching off a straight line. To obviate this the crossings are sometimes made movable. They then form a double set of facing points, and have to be treated as such; that is to say, they require to be fitted with bolts, bars and detectors. In these cases it is impossible for the bars to be placed in front of the facing point portion of the crossing, hence the bars must be placed alongside the moving portions of the crossing, and connected up as described for facing points. The point portion of the crossing is connected up like common facing points, but the two sets facing each other are coupled up to one connection from the signal box, as they both move together, but in opposite directions to set the crossing correctly for the trains to run over them (see Figs. 144, 145).

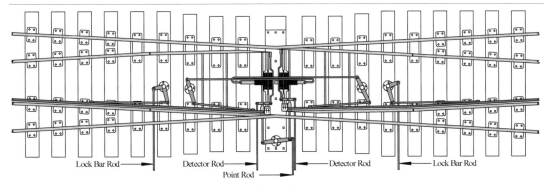

Figure 144
Movable crossing (mechanical detection)

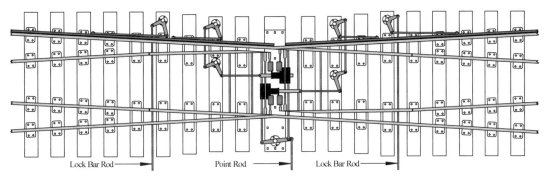

Figure 145
Movable crossing (electric detection)

COMPENSATORS

WHERE there is a run of rodding exposed to the sun or open air it continually varies in length owing to variations of temperature. Rodding exposed to the sun varies much more than rodding which is covered. Unless some means be adopted to compensate for this variation of length, it would be impossible to keep any point connections properly adjusted. For a variation of about 50 deg. F. a run of rodding 300 yards long would vary about 3 $\frac{1}{2}$ in., and when it is noted that the total travel of a pair of points is only about 3 $\frac{3}{4}$ in., it will be understood that fairly accurate compensation is necessary, so that the points may keep in correct adjustment during all changes of weather.

If two lengths of rodding are connected together by means of a straight lever, and the extreme ends of the rodding fixed, then if any variation of temperature occurs each length of rodding will vary its length to the same amount, provided (1) the two lengths of rodding are equal, (2) the temperature of both is the same, at the same moment.

The lengthening of one rod moves the lever over in one direction, but the lengthening of the other rod only assists to move the lever in the same direction, the motion of the rodding being reversed by the lever.

If now one end of the combined rods is fixed, and the other free to move, it is obvious that no variation of temperature can cause the free end to move. Should, however, the fixed end be moved, the free end will move to correspond, but in the *opposite direction* (see Fig. 146).

Any device which reverses the direction of motion in a run of rodding can be utilised as a compensator. The straight lever is the simplest device, and was exclusively used on some railway companies. It may be arranged horizontally or vertically. If it is in the form of a vertical lever, there is difficulty experienced in boxing and covering in the run of rodding, as the lever is much higher than the rods. If it is placed horizontally, it is difficult to use it in a run of rods containing many separate rods lying side by side.

Usually some arrangement of cranks to give a reversal of motion is employed, and a flat pattern is to be preferred as being the easiest to accommodate in a complicated run of rodding (Fig. 147, A). Vertical compensators usually take the form as shown in Fig. 147, C, but patterns are in use which are lowered to the same level as the rods; these give no difficulty in regard to covering; but as they require a shallow pit to receive them, there is a liability of water and mud collecting in the pit, and in frosty weather freezing solid.

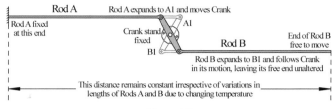

Figure 146
Temperature compensation

There is a pattern on the market using two racks and a pinion wheel (Fig. 147, D), and there is no difficulty in boxing these in, but should the teeth of the rack not be kept free from dirt, etc., friction makes them hard to work.

It is often convenient where rods have to cross the railway to arrange the crossing to be at the place required for compensators, then by fixing the cranks so that a *reversal* of motion takes place, compensation is effected without the use of a compensator, one of the cranks acting as a compensator (see Fig. 148).

In practice it is usually stated that there must be as much rod in PULL as there is in PUSH, meaning that the length of rodding under *compression* must be equal to that under *tension*.

This rule holds good for all cases of compensation so long as *each complete* length of rod compensated is *treated separately*.

It is of course essential that a compensator shall not alter the *amount* of travel imparted to the rod; it must only alter the *direction* of travel.

Where a single-ended pair of points, or a lockbar, is concerned, there is little difficulty in arranging for the compensation; only one compensator is required whatever the length of rodding

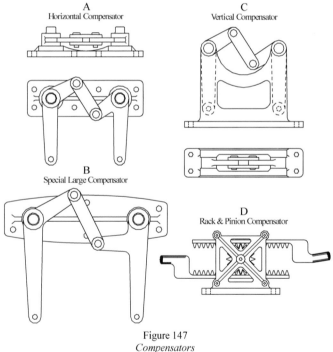

Figure 147
Compensators

(this refers to the length of rodding which can be used to work points within the Ministry of Transport limit).

In the case of a cross-over road, or any other double-ended set of points worked by one length of rod connected up to the signal box, each end of the cross-over must be treated separately.

The end which is nearer to the signal box is treated as a single point, and compensated accordingly; the rodding connecting the extreme end of the crossover road must also be treated separately, and a compensator fixed so that in this particular portion there shall be as much rod in tension as compression.

Taking a place as shown in Fig. 149, in laying out the cranks and compensators, it is desirable to commence at the signal box.

Assuming that when the lever in the signal box is pulled the rod is PULLED towards the box, the first crank will be arranged to transmit the motion unaltered. (By reversing this crank it could be made to transmit a push instead of a pull; it would than act as a part compensator, but this case will be considered later.)

Somewhere between the box and the points (near end) a compensator will be fixed, which will reverse the motion of the rodding, and at the crank marked C the motion will be *push,* the crank

C being arranged to transmit this as a *push* in order to move the point blades in the correct direction. Having determined which portions of the rod are in *push* and which are in *pull,* we can fix the correct position of the compensator; we have on the PULL side the 6 ft. rod from the first crank to the box, and on the PUSH side we have the 15 ft. length from crank C to the point blades. The total length of rod from the box to the points is 681 ft., and as one half of this must be in PULL and the other half in PUSH, dividing by 2 gives 340 ft. 6

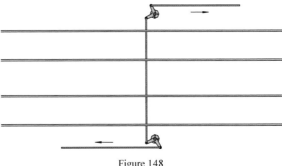

Figure 148
Crank acting as a compensator

in. as the lengths in PULL and PUSH; but we have fixed up already 6 ft. in PULL, therefore 334 ft. 6 in. of rod in PULL must be added to make up the 340 ft. 6 in. required. The position of the first compensator must therefore be 334 ft. 6 in. from the first crank. This leaves 325 ft. 6 in. (the remainder of the 660 ft. from box to points) to be in PUSH, and added to the 15 ft. already fixed as being in PUSH, completes the 340 ft. 6 in. required.

In arranging the cranks for the second portion of the crossover it will be noted that the rodding at crank C is in PUSH; somewhere between this crank and the extreme end of the cross-over a compensator which will reverse the motion must be fixed, so that at crank marked E the rodding will PULL, and as the point blades require to be pulled to set the cross-over, the crank E must be set to transmit this pull unaltered. To fix the correct position of the compensator for this portion of the rodding, it is simplest to commence at the end of the cross-over marked F.

Assume for the moment that the rodding from C to the box is disconnected.

To move the points at F (say with a crowbar), the rod between that end and crank C will be *pulled,* and its length is 15 ft. Crank C *reverses* this motion to a PUSH, which now comes to the same direction as when worked from the signal box. We have, therefore, in PULL the 15 ft. rod from crank C to the points, also the 3 ft. rod from crank E to the points, making a total of 18 ft. of PULL already fixed up, and since the total length of the rodding concerned is 198 ft., of which 99 ft. must be in PULL, therefore 81 ft. of rodding must be added to the PULL side. This is the portion of rodding from crank E to the compensator, therefore the distance from crank C to the compensator must be the remaining 99 ft. of rodding in PUSH.

Another method of arriving at the position of "D" (the second compensator), is to consider that "B" (the first compensator) is so placed as to compensate for 15 ft. beyond the crank C, so that it is possible to mark a position on the rod between that crank and compensator D which is a "neutral point"; that is, this point, 15 ft. from crank C, is unaffected by variations of temperature, therefore it is quite correct to commence from this neutral point (marked N) when fixing the position of compensator D. In this method the only piece of rod fixed up which is in pull is the 3 ft. length from crank E to the points. The total length of rod from the neutral point to the points G is 168 ft., and dividing by 2 we get 84 ft. as the amount to be in PUSH and PULL respectively; therefore, as we already have 3 ft. in PULL, 81 ft. must be added. This fixes the compensator 81 ft. from crank E and 84 ft. from the neutral point, which makes it 99 ft. from crank C, this, of course, being the same result as given by the previous method.

Taking another case, as in Fig. 150 :— Commencing at the signal box, we will assume that the first crank is made to *reverse* the motion of the rod from the box. If the rod from the box is in

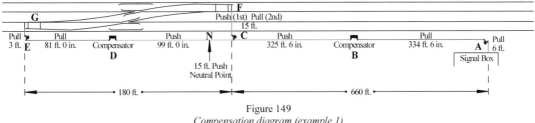

Figure 149
Compensation diagram (example 1)

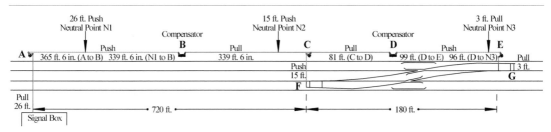

Figure 150
Compensation diagram (example 2)

PULL, the rod after crank A will be in PUSH, after compensator B it will be in PULL, and at crank C it must be transmitted to the points F as a PUSH in order to set the points correctly. After compensator D the rod will be in PUSH, but at crank E it must be transmitted as a PULL to suit the points G.

Taking the "Neutral point" method :— As crank A reverses motion there will be a neutral point 26 ft. from crank A marked N 1. Crank C also reverses motion; therefore there will be a neutral point 15 ft. from that crank towards crank A marked N 2. There are, then, two neutral points between which compensation has to be effected by compensator B. The distance between these neutral points = (26 + 15) deducted from the distance which the end F of the points is from the box, *viz.*, 720 ft., gives 679 ft., and dividing by 2 gives 339 ft. 6 in. of this rod as being in PUSH with a similar amount in PULL. Therefore compensator B must be fixed 339 ft. 6 in. from the neutral point N 1, which gives it as 365 ft. 6 in. from crank A.

For the second portion of the cross-over road, as crank E reverses motion, there will be a neutral point 3 ft. from it towards crank C, marked N 3. The compensator D has to be fixed equidistant from neutral point N 2 and neutral point N 3. The total distance between the neutral points is 192 ft., and dividing by 2 gives 96 ft. as the position of the compensator from N 2 or N 3, which is equal to 81 ft. from crank C.

In laying out rods and connecting them up to the signal box, it is usual to set the cranks on centre to allow the measurements for the lengths of the rods to be made. The positions of the compensators can be fixed on the ground by using a long piece of cord.† Commencing from the box, the cord is run out round the first crank (should it *not* reverse motion) on to the next crank, and round it to the points (should the crank there *not* reverse the motion); then the length of cord run out is doubled up, and by working back round the crank, the end of the doubled cord marks the correct position of the compensator, without any calculation.

To fix the position of the second compensator, the cord is run out from the points to the crank opposite, and if this crank reverses motion, the distance just measured is laid along the route towards the extreme end of the points, thus marking the neutral point. Then holding the end of the cord at the neutral point, it is run out round the far crank (assuming it does not reverse motion) to the extreme end of the points, then doubled up, and the end of the doubled cord marks the correct position of the second compensator, without any calculation.

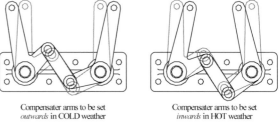

Compensater arms to be set
outwards in COLD weather

Compensater arms to be set
inwards in HOT weather

Figure 151
Setting of compensators off-centre

To measure for the length of rods all the cranks should be fixed on centre, and the measurements taken from centre to centre. At the compensators, however, some allowance should be made according to the temperature prevailing when the rods are made up. In moderate temperature the cranks of the compensators would be set on centre the same as the other cranks; in frosty weather, however, the cranks of the compensator should be set slightly *outwards*, so that when the temperature becomes normal, the cranks will be on centre with the other cranks.

---

† Chain or non-stretchable cord should be employed for *very* accurate work, but from a practical point of view a foot more or less makes no appreciable difference in the working of points.

Similarly, in very hot weather, the cranks of the compensators should be set slightly *inwards*. The amount that the compensator cranks should be set out of centre depends on the length of rod being compensated, and the temperature prevailing at the time the rods are made up (see Fig. 151). Taking the average (normal) temperature as being 60 degrees, the following table gives the amount which the arms of the compensator should be set out of centre position for different lengths of rodding and varying temperatures. The figures are given to the nearest $1/8$ in., and are based on the co-efficient of expansion for iron being 0.00000657.

ARM OF COMPENSATOR TO BE SET OUTWARDS OFF CENTRE

Temperature in Degrees Fah.	Total Length of Rodding Compensated				
	50 yards	100 yards	150 yards	200 yards	250 yards
0	$3/8$ in.	$5/8$ in.	1 in.	$1\,3/8$ in.	$1\,3/4$ in.
20	$1/4$ in.	$1/2$ in.	$5/8$ in.	$7/8$ in.	$1\,1/8$ in.
40	$1/8$ in.	$1/4$ in.	$3/8$ in.	$1/2$ in.	$1/2$ in.
60	Arms set on centre for all lengths of rod.				

ARM OF COMPENSATOR TO BE SET INWARDS OFF CENTRE

80	$1/8$ in.	$1/4$ in.	$3/8$ in.	$1/2$ in.	$1/2$ in.
100	$1/4$ in.	$1/2$ in.	$5/8$ in.	$7/8$ in.	$1\,1/8$ in.
120	$3/8$ in.	$5/8$ in.	1 in.	$1\,3/8$ in.	$1\,3/4$ in.

If some allowance is not made for the temperature when the compensators are fixed—unless they are fixed in mild weather—there is a liability of the compensators, in extreme weather, getting beyond their working limit off centre position.

In spite of every precaution in laying compensators accurately, as portions of rod are more exposed to the sun than other portions, in many cases it is impossible to obtain perfect compensation.

CONNECTING POINTS TO SIGNAL BOX

IN making the connection to the box complete, it is usual to arrange the travel of the rod as it leaves the box to be about 5 in.

The cranks between the box and the near end of a double-ended point connection should be equal-armed cranks, so that this 5 in. is unaltered down to the last crank at the points. There will be some loss of travel depending on the fit of the joint pins, and the rigidity, or otherwise, of the rods. The crank arms at the points should be made to reduce or increase the travel to suit the points. For ordinary points the travel required is about 3 ¾ in., and between the last crank and the points there is an adjusting screw, by means of which the points are regulated exactly, by tightening or slackening it out as required. There should be the same amount of tightness, or "nip," between the point blade and the stock rail, when the points are in the normal position, as when they are reversed. When the ends of the point blades, which are nearer to the box, have been properly adjusted, the far end is connected up. The final crank at this end will require to give a slightly greater movement than the crank at the nearer end, as there will be a loss of travel owing to the additional joint pins, and additional length of rodding involved.

If adjusting cranks are used there is no difficulty in arranging the travel to suit. In the case of wrought-iron cranks with two holes in each arm, it is a matter of putting the joint either in the inner or outer hole; with cast-iron cranks, a different size of crank has to be used, with length of arm to suit the requirement.

It is very necessary to see that the points are travelling the correct amount, and should the gauge be slightly tight or slack at one end of the points, it is very difficult to make both ends work together properly.

COVERING

WHEN point rods are fixed in such a position that they are liable to form an obstruction to shunters, etc. (to comply with the orders issued in virtue of the Prevention of Accidents Act), they

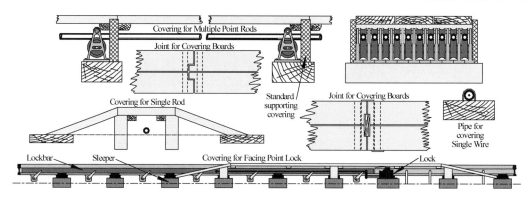

Figure 152
Wooden covering for point rods and signal wires

must be covered over or boxed in. Fig. 152 shows various types of coverings used. Where a run of rodding requires covering it is advisable to keep the sides as open as possible for free ventilation, otherwise the rods rust up very quickly, and they are not so easy of inspection.

Facing point gear and point connecting rods are usually covered over with detachable coverings. Wood in most cases is used for this work, but sometimes iron is used for facing point covering in the 4 ft. way.

Paragraph 6 of the Ministry of Transport Requirements refers to Points and reads as follows :

POINTS

6. Points to be so situated that movements over them shall be within view of the signal box from which they are worked, unless an approved alternative for direct vision by the signalman, e.g., track circuited diagram, is provided.

The limit of distance from levers working points to be 350 yards, unless the points are power worked and occupation of the lines is electrically indicated in the signal box, in which case the distance may be indefinitely extended.

All points to be fitted with not less than two stretchers. Rodding, or other approved method, is to be used throughout for the mechanical working of all points, and also for bolting them when required. In certain conditions the use of unworked trailing points will be permitted.

Facing points on passenger lines, and all points commonly used in the facing direction by passenger trains, to befitted :

a) With bolt-lock through a third stretcher; and with locking bar, or some other approved device. The operation of the bolt-lock must depend upon the correct movement of the locking bar where it is used. The length of locking bars to exceed the greatest interval between any two adjacent axles likely to be used on the line.

Where approach track circuiting is used in substitution for the locking bar at power-worked facing points, the controlling track should, wherever possible, be of sufficient length, having regard to speed of traffic and time taken to operate the points, to ensure that, if the point lever is worked immediately before the track circuit is occupied, the points shall have completed their movement before the train reaches them.

b) With stock rail gauge tie.

c) With means for detecting the position of each switch, as well as the bolt-lock by the relative signals.

It is desirable that:

i) All trailing points used in the facing direction for shunting movements should be detected with the relative signals. Single switch detection for each direction of movement will be accepted in such instances.

ii) On goods lines, used exclusively for running movements, facing points should be equipped as on passenger lines.

6
INTERLOCKING APPARATUS

To enable the signalman to operate the points and signals from his box, levers are arranged in a row, to which the point rodding and signal wires are connected, the levers being interlocked to ensure that the signalman shall not clear conflicting signals, or have points and signals pulled together which could lead to a collision. The whole apparatus is commonly termed a "Locking Frame." This includes the levers, the interlocking, and the supporting framings.

There are many types of locking frames in use; these differ greatly in principle of operation and in design. The shape of the levers varies considerably, as does the main framing and locks. The levers, as a rule, are made to stand approximately vertical in the signal box when in the NORMAL position; when the levers are pulled they move through an arc of a circle. In some designs the lever is simply a straight piece of metal with a handle formed at the top; the lower portion of the lever is fitted into a bearing plate or "shoe" (Fig. 153). In most cases the connections from outside to the levers are led upwards, so that a "tail" is required at right angles to the lever for connecting purposes. The lever and the tail may be separate pieces, each bolted to the bearing plate. In other designs the lever is made in one piece and bent to form the "tail," and sometimes set so as to make the portion above floor level assume a vertical position when NORMAL.

The bearing plates are made of cast iron, and have pins fitted to them, on which the lever works. The pins in turn are supported on cast-iron sole plates (Fig. 154, A). With this arrangement it is possible to unbolt the lever and take it out without disturbing any other portion of the locking frame, a very valuable arrangement when alterations have to be made while the frame is in service.

The length of tail can be altered to suit the particular purpose required, being about 18 in. for working points and about 2 ft. 3 in. for working a signal. It is sometimes arranged that the bearing plate accommodates a "back tail" with a weight attached, which assists in pulling off a signal when the signal is a long distance from the box.

The bearings are placed about 3 ft. to 4 ft. below floor level, and the lever top is about 3 ft. 3 in. to 3 ft. 6 in. above floor level. To keep the levers apart, fill up the space between the levers, and at the same time to supply a "stopping" point at floor level, floor plates are bolted to angles, which in turn are fastened to upright supporting frames. Each floor plate can be taken off separately in case of renewal. In

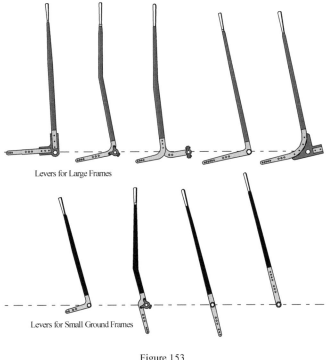

Levers for Large Frames

Levers for Small Ground Frames

Figure 153
Levers for locking frames

some early patterns the levers are threaded on one long shaft, with distance collars to keep them apart, and the floor plates are threaded on a long through bolt, which extends the whole length of the frame, binding the supporting frames and floor plates together. With this type some difficulty is experienced in taking out a defective casting for renewal. It is very essential that any portion of a frame shall be capable of being taken out and replaced without interfering with any other portion more than is absolutely necessary.

When the lever is all in one piece a clip bearing is bolted on (Fig. 154, B), so that the lever can be taken out, or the bearing tightened up when worn, without interfering with any other lever. The clip is fitted to a shaft with recesses turned in it (the clip being in the recess), the full-sized portion of the shaft serving as a distance

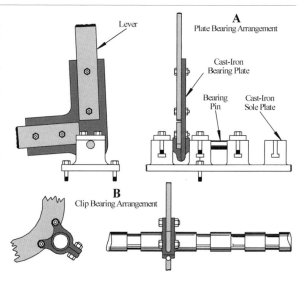

Figure 154
Bearings for levers

collar. The shaft is threaded through the supporting frames and held secure by set pins. The shaft does not enter more than half-way through the end supporting frames, so that additional levers may be added without interfering with the existing portion of the frame. The recesses in the shaft fix the pitch of the levers, the usual pitch being either 4 in. or 5 in., although some patterns have 4 1/4 in. and 5 1/4 in. pitch. With a solid pattern lever it is not possible to change the length of the tail, so, to save any necessity for changing levers in case of an alteration, all the lever tails are, as a rule, made long enough to suit signals, but holes are drilled further back to suit points. There are several holes drilled in each tail to allow of the outdoor connections being regulated. The usual travel for a signal lever at the hole in the tail is about 7 in. for the hole nearest the lever bearing and about 12 in. from the furthest out hole. In the case of a point lever it is about 4 1/2 in.

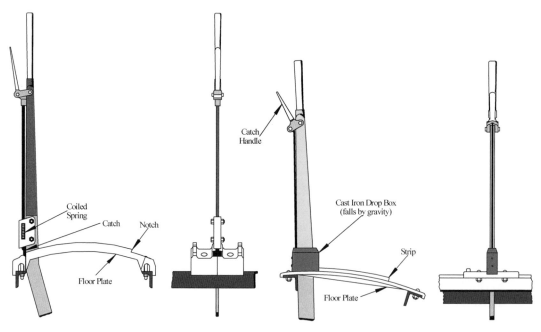

Figure 155
Catch handle mechanisms

for the nearest hole, and about 6 $\frac{1}{2}$ in. for the furthest out hole.

To hold the levers in either the NORMAL or OVER position, catches are fitted to the levers. The catch engages in a projection on the floor plate and is operated by a "catch handle" fitted near the top of the lever (Fig. 155). The catch may engage with the notch or strip of the floor plate, either by the help of a spring or by gravity alone. If gravity only is used, a heavy cast-iron "drop box" is fitted to the lever. In some old frames there were several notches in the floor plate for a signal lever, allowing it to be pulled varying amounts, according as the signal might work easily or not, and the travel

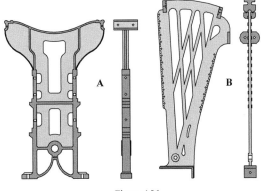

Figure 156
Supporting frames

of a point lever was less than the travel of a signal lever. In most modern frames, however, all the levers move through the same distance. The mechanical advantage in the case of a point lever is about 6 to 1, and about 3 $\frac{1}{2}$ to 1 as an average for signals. In pulling a lever the hand moves, in the case of older types of frame, about 30 in., strokes up to 48 in. being provided in newer types.

FRAMES

THE supporting frames, called "standards," or "cheeks," are made of cast iron, usually of open design for lightness, and ribbed for strength (Fig. 156). Feet are cast on the standards to allow of their being bolted to a girder or beam, and arrangements have to be made for accommodating the bearings of the levers. The angle irons to which the floor plates are bolted and the locking must also be supported by the frame.

The frames should be designed so that where an extension is being made additional levers can be inserted without necessitating the removal of the end standards of the old frame. It is usual to place a standard after every tenth lever, so that in a 30-lever frame there would be four standards, one at each end, and two intermediate. The spacing of these standards depends on the class of frame. Where the levers are pitched 6 in. apart, and in some patterns where the pitch of the levers is 5 $\frac{1}{4}$ in., the standards are placed after each set of eight levers. It is desirable to have the spacing of the standards invariable, as it allows of fewer castings being stocked for repair work. In making up a frame it often involves too many spare levers to fix on some multiple of eight or ten as the total number of levers in the frame, so it is generally arranged that the end standard may be placed after the last four, where eight is the usual number, and the last five where ten is the usual number. Thus castings have only to be stocked for a complete bay of eight (or ten) levers, and for half a bay of four (or five) levers.

HOOK LOCKING

IN most modern looking frames the actual interlocking of the levers is effected on the "TAPPET" or "WEDGE" principle. This is a development of an early type of locking frame known as HOOK

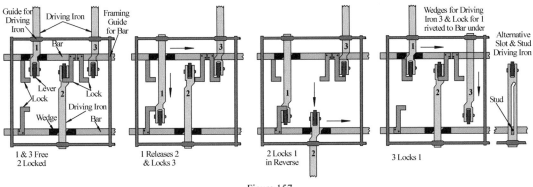

Figure 157
Hook locking

LOCKING. In this type of locking, flat pieces of iron were fitted to each lever. These were shaped with a bevel, which bevel engaged in a wedge riveted to a long bar (see Fig. 157). The bar was supported at each frame standard, and was guided so that it could only move longitudinally. In pulling over the lever its driving iron moved the long bar. To this bar were riveted hooks which engaged with the back of the levers, preventing their being pulled unless the hooks were moved clear first. The bars carrying the hooks were placed one below the other; about twenty could be accommodated at the back of the frame and about ten at the front of the frame. The hooks at the back of the frame were for the NORMAL locking, and the hooks at the front were for OVER (or reversed) locking.

TAPPET OR WEDGE LOCKING

IN developing the WEDGE locking, the driving iron is retained, having notches cut in it which engage into a wedge piece as in the hook locking, but instead of the wedge being riveted to a long bar it is formed on the end of a small rod. The bar is retained, but instead of being capable of

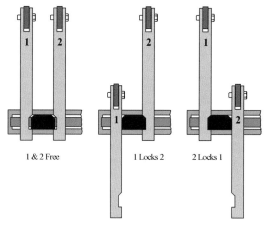

Figure 158
Locking example (1 locks 2, 2 locks 1)

moving longitudinally it is fixed. Guide pieces are riveted to this bar, which hold the driving irons and the wedge-shaped rods in position. When a lever is pulled, the driving iron forces the wedge out of the notch and the wedge rod moves along; the opposite end of the wedge rod has a similar bevel formed on it, which enters into a notch cut in another driving iron belonging to some other lever, so that when the first lever has been pulled—the wedge being in the notch of the second lever's driving iron—it is not possible to pull that lever. By arranging the notches in the driving irons and the wedges on the ends of the rods properly, it is possible to give effect to any desired combination of locking. The usual items of locking required are (1) one lever to LOCK another; (2) one lever to be RELEASED by another; and (3) one lever to LOCK another lever BOTH WAYS (normal and reversed). In addition to this there is *special* or *conditional* locking, where one lever locks or releases a second lever only when a third lever is normal or reversed.

Fig. 158 shows lever 1 locking lever 2. It will be noticed that lever 2 in turn locks lever 1, and in any locking arrangement, if lever A locks lever B, it follows that in return lever B *must* lock lever A. In some designs of locking frames (hook locking, for instance), the mere insertion of A locking B does *not* of itself ensure that B locks A, and the return locking requires to be inserted *separately*. With wedge locking the return locking never needs to be inserted separately; the return locking is always provided automatically.

Fig. 159 shows lever 1 released by lever 2—that is, until 2 has been pulled it is impossible to pull 1. The *return* locking in this case means that when 2 has been pulled, and 1 also pulled, it is *impossible* to put back 2 until 1 has first been put back. Lever 2 is now said to be BACK-LOCKED, or *locked in the reversed position*.

Fig. 160 shows lever 1 locking lever 2 both ways. If 1 is pulled when 2 is normal, it is then impossible to pull 2, and 2 is therefore locked normally. If 1 is now put back, and 2 pulled, it is

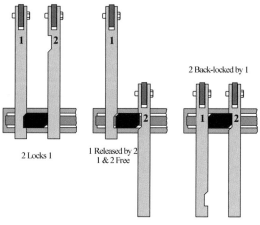

Figure 159
Locking example (1 released by 2, 2 back-locked by 1)

possible to pull 1 (unlike the case shown in Fig. 158), but if an attempt is made to put back 2 whilst 1 is over, 2 will be found to be BACK-LOCKED. It therefore follows that when 1 is normal, 2 can be worked backwards and forwards, but as soon as 1 is pulled, 2 is locked, in whichever position it happens to be at the moment. It should be noted that unless 2 is either correctly NORMAL or REVERSED, 1 cannot be moved; this is sometimes referred to as "2 locking 1 *during stroke*."

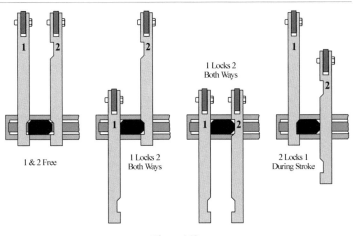

1 & 2 Free

1 Locks 2 Both Ways

1 Locks 2 Both Ways

2 Locks 1 During Stroke

Figure 160
Locking example (1 locks 2 both ways, 2 locks 1 during stroke)

SPECIAL LOCKING

SPECIAL locking is effected by allowing one or more driving irons to move laterally, instead of being held in line by the strips of the bar. To prevent the driving iron from sliding when not required, a piece of lock rod butts against it, holding it in the same position as the strip on the bar would. Fig. 161 shows lever 1 released by 2, and locking 3 when 4 is pulled. There are several ways of giving effect to this, either 1 or 4 could be made to slide, or it could be arranged for both 2 and 3 to slide, as in Fig. 161, C. There are numberless possible combinations of special locking, all worked on the principle of a driving iron being allowed to slide away from a lock under certain conditions fulfilled by other driving irons.

In this pattern of locking, should more than one bar be required, they are placed one below the other, and it is possible to have 13 bars in the back and 8 in the front of the frame. The wedge locks are connected together by short bridge pieces; these are used to pass over one or two driving irons, and it is general to use stronger rods to connect the locks when the rods are required to pass over several driving irons, or over the bridge pieces. Short connections can also be made by slotting the bar and placing the connecting piece in the slot. The locking pieces and driving irons are prevented from being displaced by means of covers screwed to the guide strips, and all bridge pieces and lock connecting rods have to pass through slots cut in these covers (see Fig. 163). Driving irons fitted low down on the lever have a very short travel; consequently the signalman has a very powerful leverage to force the locking. Should a frame be very large, sufficient locking cannot be placed in single bars as described above; double bars are then resorted to, and are fixed in the same manner as single bars (see Fig. 162). Instead of only one locking rod or lock connecting rod being fitted to each bar, two rows are fitted. Where two wedges are fitted to one driving iron on the same side of it, care must be taken that the wedges will not enter the wrong notch, and so lock the lever before it is in the correct normal or reversed position. Various methods are adopted to prevent the wedges entering the wrong notch, the most common being to

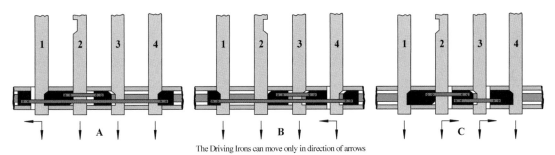

A   B   C

The Driving Irons can move only in direction of arrows

Figure 161
Various arrangements of driving irons to effect 1 released by 2, and locks 3 when 4 is pulled.

make one of the wedges larger than the other, so that the large wedge will not enter the small notch (see Fig. 164). Sometimes as many as three rows of locks have to be employed, but this is inconvenient, as greater care needs to be taken to ensure that there shall be no locks entering wrong notches.

In a more modern type of this locking frame the "bars" are of cast iron. All guiding strips in this case are cast solid with the other parts of the bar. The bars are cast in lengths to suit the spacing of the supporting standards, and are carried on short brackets bolted to the standards. In the cast-iron bar a channel is formed in the bottom to accommodate a bottom connecting rod; this takes the place of the slot required in the wrought-iron bar. Sheet-iron covers are fastened on to the bar by set pins, or bolts and nuts, and all top connecting rods have to pass through slots in this cover. The locking wedges are small pieces riveted to flat connecting rods and lie in the bottom of the channel. Top connecting rods are avoided as much as possible, and unless they are arranged so as to be in tension, should be as short as possible. In the cast-iron "bars," channels are cast to accommodate two rows of locks, and if a driving iron is required to slide laterally for conditional locking purposes, it can be made $1/2$ in. narrower than the space provided in the casting, and so avoid having to widen the space in the casting.

With the bars of locking set one below the other, it is rather difficult to make an examination of the frame without stripping a large quantity of the locking, and before a cover can be taken off the top connecting rods have to be removed, thus disconnecting some of the locking.

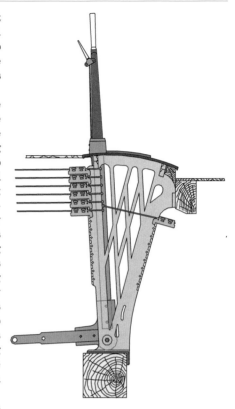

Figure 162
Stevens & Sons' pattern frame

A further development of this class of locking frame places the locking bars in steps (as in the N. E. R. Pattern Frame, Fig. 165). This allows any bar to be examined without much difficulty. In this pattern the cast-iron bars are much deeper, so that the top cover encloses all the connecting rods and locking; this allows a cover to be taken off without disarranging any of the locking. With the deeper bars it follows that there cannot be quite so many of them; so to allow of the same amount of locking being inserted in a frame the bars are designed to take more connecting rods. With bars of this size they are usually termed "troughs," or "locking boxes." As a rule only two

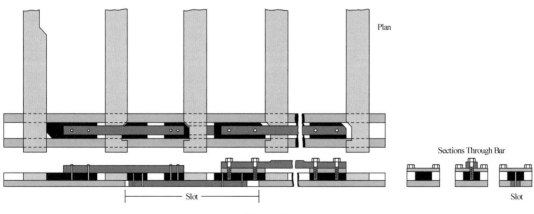

Figure 163
One-channel bar

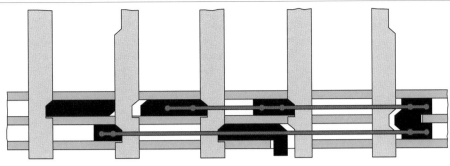

Figure 164
Two-channel bar

rows of locks are allowed, but each channel can take five connecting rods, so that a complete locking box would take ten lock connecting rods, two in each channel being placed below the locks and three in each channel being placed above the locks (see Fig. 166). The locks are of different shape for each channel to avoid conflicting locks and notches.

Various names are used for the "driving irons"; they are sometimes called TAPPETS, PLUNGERS, or SWORD IRONS. The locks are sometimes termed WEDGES, DIES, or TAPPETS. The "connecting rods" are sometimes called BRIDLES or BARS.

The most convenient arrangement of locking is to have the channels placed flat like a table, but if slightly inclined the locking is easier to inspect. When the driving irons are worked directly off the lever, as in the above designs, it is difficult to arrange for more than three channels in one plane without very carefully spacing the notches to prevent wrong locks entering. If the travel of the driving iron is so short that one notch never moves sufficiently far to allow of its coming opposite a lock in another channel, there can be no difficulty about the locks entering wrong notches. This

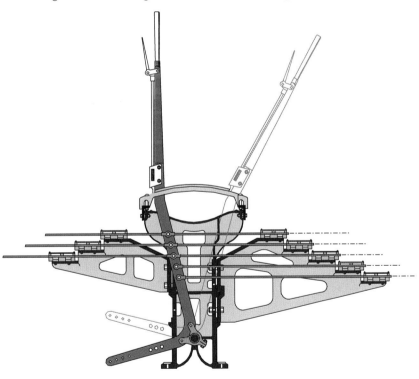

Figure 165
North Eastern Railway (N. E. R.) pattern frame

could be arranged by having very long driving irons, and pitching the channels further apart than the travel of the driving iron. If the travel of the driving iron is made very short, unless some travel reducing gearing is adopted, it must be placed too low down on the lever to be of any use for locking purposes. In several designs of locking frames the travel of the driving iron has been reduced to about 2 $\frac{1}{4}$ in., and the pitch of the channels fixed at about 3 in.

CAM AND CAM SLOTTED PLATE MOTION

ONE of the best-known devices for reducing travel is that of the cam (McKenzie and Holland's Frame, Fig. 167). The cam is operated by means of a stud fixed to the side of the lever; the first

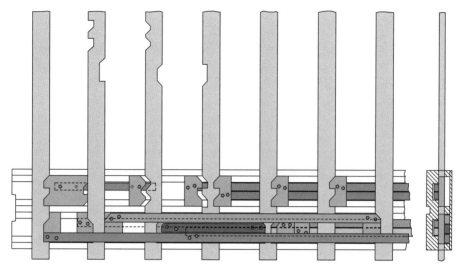

Figure 166
Cast-iron locking troughs

movement of the lever moves the cam slightly and displaces the driving iron about 1 1/8 in. The stud on the lever now slides along a radial portion of the cam, and this imparts no motion to the driving iron, but just before the stop on the floor plate is reached the stud again comes to a non-radial portion of the cam, and another 1 1/8 in. travel is imparted to the driving iron. This is sufficient to ensure an effective NORMAL and REVERSE lock.

With this pattern there is only one driving iron (called a SWORD IRON or PLUNGER); which can be of any required length. Five connecting rods are placed in a channel if required, and all the looks (TAPPETS being the term used) are V-shaped, there being no conflicting notches in the sword irons (see Fig. 168).

Another very similar device is in use for reducing the motion of the locking plungers, as shown in Fig. 169 (G. W. R. Pattern Frame).

In this pattern, instead of a cam working on a bearing, a plate with a cam-shaped slot cut in it is driven by the lever. The cam slot imparts its motion to a vertical member, which in turn is connected to cranks, and to these the locking plungers are attached; or the

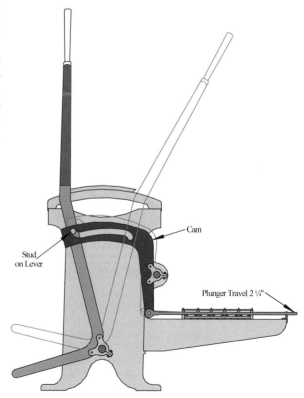

Figure 167
McKenzie & Holland's cam and tappet frame

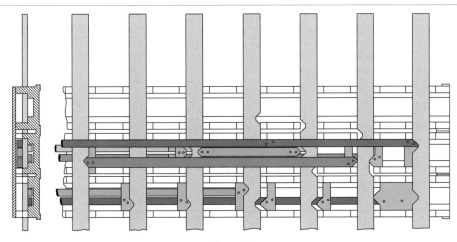

Figure 168
Table locking—travel of plungers, 2 $\frac{1}{4}$ in.

locking plungers may be connected directly to the vertical slide and arranged in line therewith.

RACK AND PINION MOTION

ANOTHER method employed is shown in Fig. 170 (Sykes' Pattern Frame). There are two racks side by side, one stationary, the other movable and connected to the locking plunger. On the lever a small pinion is fitted, and this pinion engages with both racks during a portion of its travel. When the lever is normal the pinion only engages with the rack fitted to the locking plunger—the fixed rack being cut away—and is prevented from rotating by a fixed tooth which engages with it. When the lever is pulled the pinion, being unable to rotate, moves the rack with which it is in mesh, and so operates the locking plunger. When the lever has moved a short distance, the pinion has moved clear of the fixed tooth and is now free to rotate. The pinion having arrived at this position is just engaging with the fixed rack, and this prevents the locking plunger from moving out of its position while the pinion is running across both racks. Before the end of the stroke the fixed rack is again cut away, and the pinion engages with another fixed tooth, thus imparting a final motion to the driving plunger. In putting the lever back to the normal position the reverse action takes place.

CONDITIONAL AND ROTATION LOCKING

WITH any of the pattern locking frames which only use *one* locking plunger per lever, it is obvious that it is impossible to allow the plunger to be used as a sliding plunger for conditional locking purposes, otherwise it would not be possible to insert any more locking for that particular lever.

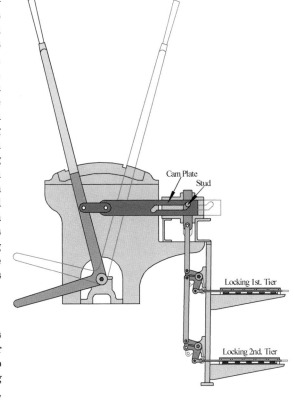

Figure 169
G. W. R. device for reducing travel of plunger

Where conditional locking has to be put in, some form of sliding or hinged plate is used. A short plate is fitted to the locking plunger, this plate sliding or swinging laterally instead of the locking plunger (see Fig. 171).

Another method is to employ a special L swinging piece which is fitted to the lock (Fig. 172, A). Either of the connecting rods shown can move when the lever is pulled, thus allowing any conditional locking to be set out. Also a tongue piece can be fitted to a lock, the tongue pressing against two connecting rods either of which can move away, when the lever is pulled, to effect the conditional locking (Fig. 172, B).

It is occasionally necessary to insert locking to prevent a home signal from being pulled a second time, unless the starting signal has been pulled and put back again, locking of this description being termed "rotation" locking.

Fig. 173 shows one method of carrying out this type of locking, and illustrates the case of lever 3, which, having been pulled and put back, cannot be pulled a second time until *either* 1 or 2 has been pulled and put back.

The plungers on levers 1 and 2 are similar. When either of these levers is pulled the plunger rises over the lock, owing to the bevel on the lock and corresponding bevel on the plunger; and when the lever is pulled over far enough the plunger drops down to its normal plane by gravity alone.

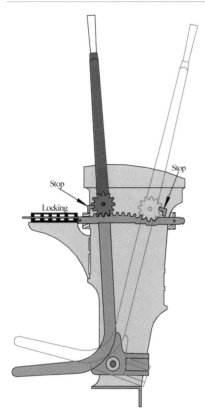

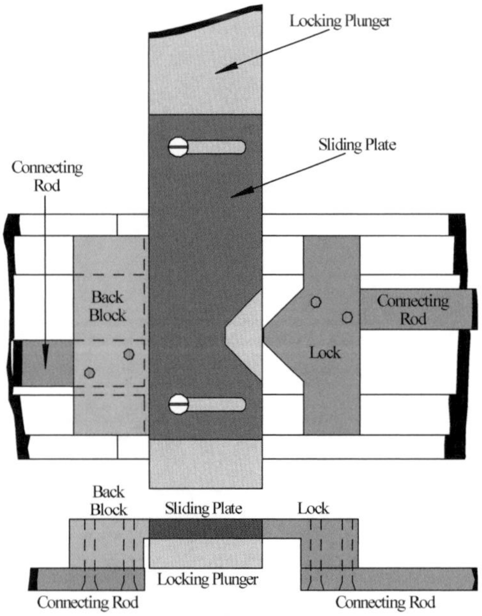

Figure 170
Sykes' rack and pinion device for reducing travel of plunger

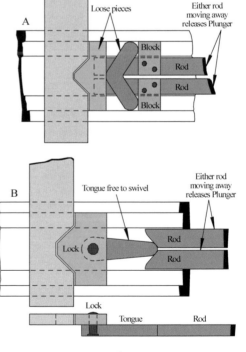

Figure 171
Sliding plate special locking

Figure 172
Swing tongue special locking

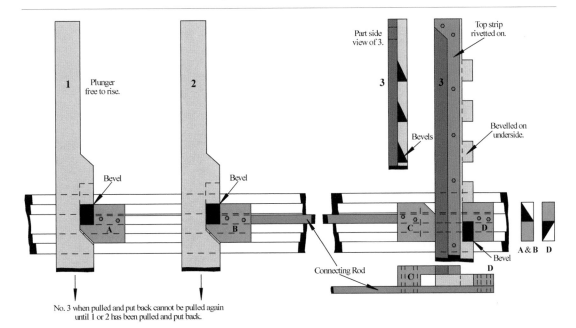

Figure 173
Rotation locking

When the lever is being put back again to its normal position, the plunger forces the lock along to the right—not being able to ride over the lock, as when it moves in the reverse direction—and this movement of the lock is transmitted by the connecting rod to locks C and D, moving D clear of the plunger on lever No. 3. This, then, allows No. 3 to be pulled. When 3 is pulled right over, the bevelled portion of the top strip engages with lock C and returns the locks to their original position. When No. 3 lever is put back, the plunger rises over lock D owing to the arrangement of the bevels, but as soon as the plunger has moved the first notch over lock D it is impossible to pull it again, and the lever can only be moved in the direction of restoring it to the normal position.

The lever having been put right back it remains locked until either 1 or 2 is again pulled and put back, when it is once more released.

In the case illustrated No. 3 would be an outer home signal, and 1 and 2 would be junction signals ahead. If a train is to be sent to stand at either of the junction signals, 3 would be pulled to allow it to draw forward to 1 and 2. On 3 being put back to danger it is impossible for it to be pulled again to send a second train to stand at the junction signals, until either one or the other of the junction signals has been lowered to let the first train away, and then put back to danger again to hold the second train.

This class of locking is not commonly fitted, except where the traffic is difficult to operate, because it needs more than the usual amount of attention to maintain it in efficient working order.

CATCH HANDLE LOCKING

THE foregoing pattern locking frames are sometimes termed LEVER LOCKING frames, to distinguish them from another class known as CATCH HANDLE LOCKING frames. In the lever locking frames the pulling of the lever operates the locking, while with the catch handle locking frames the movement of the catch operates the locking.

The majority of locking frames in this country arc of the former type. In theory it is far preferable that there should be a preliminary lock before the lever can be pulled. There have been instances of signalmen, when unable to pull a lever, *assuming* that the locking was at fault, and waving the train past the protecting signal with disastrous results. With catch handle locking no such mistake can occur, as, if the catch can be lifted clear and the lever still cannot be moved, it is certain that the outdoor connections are at fault; on the other hand, if the catch cannot be raised, it is certain that the interlocking is holding the lever. It is obvious that if the catch cannot be lifted it is impossible for the lever to be pulled.

The earliest type of catch handle locking used a slotted plate (Saxby and Farmer's Frame, Fig. 174). This is fitted to the floor plate and engages with a stud fitted to the catch rod. When the catch is lifted it moves the plate slightly, and the lifting of the catch allows the lever to be pulled; while the lever is being pulled no motion is imparted to the slot plate, as the slot is cut to the same radius as the motion of the stud. When the lever is right over the catch handle can be dropped, allowing the catch to enter into the notch in the floor plate. This, then, gives another small motion to the slotted plate, but as the stud is on the opposite side of the centre pin on which the slotted plate works, the latter moves in the same direction as on the first movement. It is important to notice this, as, should the final motion of the catch handle *reverse* the initial motion, the apparatus would be useless; for, although it would still be possible to LOCK a lever either in the "normal" or "reversed" position, it would be impossible for the locking to ensure that the lever concerned is reversed when required, or normal when required. It is useless to

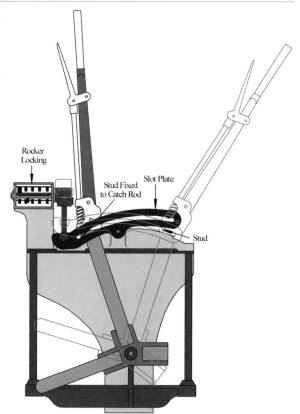

Figure 174
Saxby and Farmer's rocker locking frame

have a lever locked in the normal position when it should only be locked in the reversed position. In all catch handle locking frames some device must be employed *to register the actual position* of the lever. The slot plate in the above device was attached to rocking shafts which, in turn, were connected to long rods, and to the latter locks were fitted. On the rocking shafts grids were formed, and the locking was effected by arranging that the locks prevented the grids from rotating, or the grids prevented the locks from travelling, when certain levers were, or were not, pulled (Fig. 175).

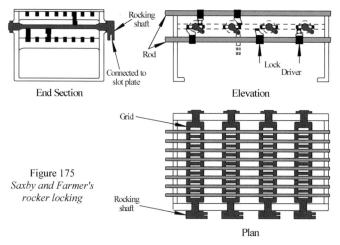

Figure 175
Saxby and Farmer's rocker locking

A modern development of the above interlocking frame employs a form of wedge locking. The slot plate is retained and connected to rocking shafts, the rocking shafts having short driving pinions fastened to them, and the pinions engaging with teeth cut in the long bars which extend along the frame. To these bars wedge-shaped pieces (called "dogs") are screwed, which engage with cross-locks to effect the interlocking.

This pattern locking frame is largely used for manual installations in the United States of America, and a modification of it is also used to effect the mechanical interlocking on some of the Power Signalling Installations adopted in this country (Fig. 176).

"Special" locking in this improved wedge-locking pattern is effected by allowing a "dog" to swivel on a pin-bearing, instead of being screwed on to the bars rigidly.

Another type used in this country to some extent is the DOUBLE PLUNGER catch handle locking frame (Saxby and Farmer's, Fig. 177). Two driving plungers are fitted one above the other in the same locking channels. The bottom one is

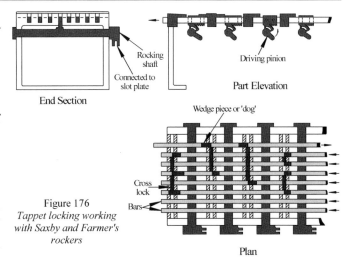

Figure 176
Tappet locking working with Saxby and Farmer's rockers

connected directly to the lever, but very low down; consequently, its travel being small, it alone would not be strong enough to give an effective lock. The top driving plunger is connected to the catch handle, and this plunger first drives the locking; if the lock cannot move, the catch cannot be lifted, hence the lever is locked. This plunger does all the locking, both reversed and normal, the bottom plunger, which is connected direct on to the lever, being for the purpose of *registering the correct position of the lever*. The catch-operated top plunger in some portion of its movements occupies the same position whether the lever is normal or reversed, and without the registering plunger it would be possible for the catch handle to be locked in a wrong position.

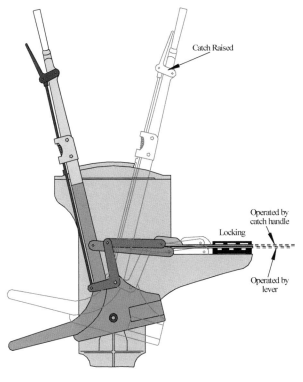

Figure 177
Saxby and Farmer's double plunger locking frame

Fig. 178 shows another type of catch handle locking in which only one locking plunger is used—Dutton's pattern— the motion being imparted from the catch rod by means of a slot and sliding pin to the locking plunger. When the lever is normal the slot is inclined at an angle to the locking plunger, so that when the catch handle is lifted the connecting pin slides down the slot, and in so doing moves *towards* the locking channels, carrying the locking plunger with it. This portion of the travel gives effect to all NORMAL locking. The catch having been raised, it is possible to pull the lever over, and, in doing so, the pin remains at the bottom of the slot, but as the bottom of the slot is not quite on the same centre line as the lever centre, a slight motion is imparted to the pin, and thus to the locking plunger. This has no effect on the locking, the motion being so arranged that the locking plunger travels slightly towards the NORMAL position, but *not* far enough to take up the normal position, so that the notches in the locking plunger are still out of register with the locks. When the lever is fully reversed the slot is again inclined at an angle to the locking plunger, but this time the inclination is *reversed*, as the *top* of the slot is nearer to the locking channels than is the bottom of the slot. When the lever is normal the *bottom* of the slot is nearer the locking channels than is the top of the slot, and thus, when the catch is dropped in the notch, the pin, in moving *up* the slot, forces the locking plunger further through the locking channels, so giving effect to all the REVERSE locking. Fig. 179 shows the various positions of the lever and locking plunger. The locking channels in this pattern are inclined to about 45 degrees, and this makes it very convenient for examination or alterations. All the locking is held down irrespective of the top covers, which are solely for the purpose of keeping out dirt and dust.

In most catch handle locking frames one portion of the stroke of the locking plunger depends on the spring which forces the catch into the notch. This necessitates very strong springs, as should a spring not force the catch right home the locking would be fouled. This makes the levers rather tiring to work, as the grip of the handle not only has to compress this spring, but also has to drive all the locking, and with large frames this trouble becomes acute. With catch handle locking all the locking parts can be made very much smaller, and consequently lighter than is the case with locking working directly off the lever. It is common to screw all locks on to the connecting rods; and this is very convenient when it is necessary to take off a lock, as it can be unscrewed without interfering with much of the other locking; but screws are not very convenient when it comes to making alterations on the ground. Very often locking alterations of a very extensive nature have to be carried out within twenty-four hours, and when connecting rods have to be drilled and then tapped for a screw a considerable amount of time is occupied, also, unless the holes are drilled very accurately, great difficulty is experienced in tightening the screws right home. Locks which are riveted to the connecting rods are much easier to deal with, as any slight inaccuracy in drilling can be ignored when driving the rivet in.

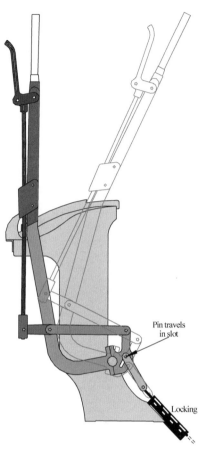

Pin travels
in slot

Locking

Figure 178
Dutton's catch handle locking frame

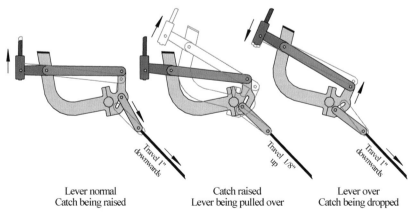

| Lever normal | Catch raised | Lever over |
| Catch being raised | Lever being pulled over | Catch being dropped |

Figure 179
Relative motions of lever and plunger, Dutton's locking

In one pattern catch handle locking frame (L. & N. W. R. pattern, Fig. 180), the handle is placed at right angles to the lever, and no spring is used; the signalman depresses the handle to raise the catch, and when the lever is over, the catch handle is raised to put the catch in the notch.

The catch takes the form of a cast-iron drop box, which engages with a radius bar on the floor plate.

From the drop box a connection is taken to a small crank, which is centered on the casting which acts as a bearing for the lever. A rack slide with a specially shaped jaw fits round this crank, and when the lever is normal the end of the crank to which the catch rod is connected has a stud which engages with the jaw on the slide. When the drop box is raised, it pulls the crank *up*, and the jaw slide consequently moves *up* with it. As the lever is being reversed the crank travels through the jaw; and when the lever is fully reversed the opposite end of the crank engages with the jaw, so that when the drop box is lowered a further upward motion is imparted to the slide. The interlocking proper is fitted to the rack sides. Fig. 182 shows the positions of the lever and locking slide.

FITTING OF LOCKING

IN fitting locks a slight amount of play should be allowed between the notch in the plunger and the lock, and with locking working directly off the lever this should be sufficient to allow the lever to be pulled over about $^1/_4$ in., or slightly more. It is important that the slack shall *not* be between the locks and the driving plungers in a

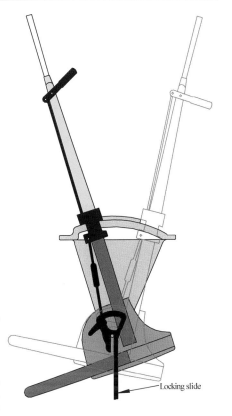

Locking slide

Figure 180
L. & N. W. R. catch handle locking frame

longitudinal direction. The locking should be kept as tight as is consistent with easy working in this direction, and the slack should only be in the *notch* of the plunger. If locking is fitted with the notches too tight when the frame is in the shops, it will be found, on its being erected and connected up to points and signals outside, that the point levers all tend to be forced back against the *notch* of the floor plate, while the signal levers all tend to lie tight against the *stop* of the floor plate. There must be some clearance between the notch and the catch for easy working, and this difference is sufficient to make the locking plungers move enough to foul the locks if the locking is fitted absolutely tight.

With tight locks it is also difficult to know whether the lever is stuck or not; the lever should move quite easily until it is checked *dead* by the locking. Springy locks should not be allowed, as

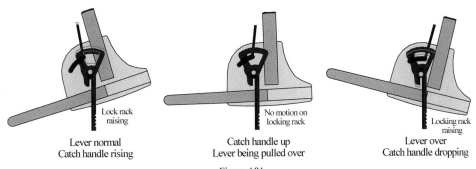

Lock rack raising

Lever normal
Catch handle rising

No motion on locking rack

Catch handle up
Lever being pulled over

Locking rack raising

Lever over
Catch handle dropping

Figure 181
Relative positions of lever and locking slide, L. & N. W. R. catch handle locking frame

they indicate that some of the connecting rods are tending to buckle, or that some of the parts are not tight.

When the locking is operated by means of a cam, or other reducing motion, the amount of slack admissible is rather *less* than $1/4$ in., and with catch handle locking the slack should be about $1/8$ in.

The chief features required to make an efficient locking frame may be summarised as follows:

(1) Few working parts, *especially* as few pins between the lever (or catch handle) and the locking, as possible. Each pin means wear, and eventually slack locking, together with the risk of a pin dropping out and disarranging the locking.

(2) Locking easily accessible for inspection, cleaning, and renewal.

(3) Wearing parts to have broad bearing surfaces, and the locks strong and effective.

(4) Should not add much additional labour to the signalmen in working the outdoor connections.

(5) Should be simple to understand and easily put together, or taken apart.

(6) All levers interchangeable, and the framing arranged so that extensions can be made easily.

(7) The amount of locking which it is possible to insert in a frame should not be limited for constructional reasons, and the type should be suitable for any number up to about 200 levers.

Possibly no one interlocking apparatus can lay claim to all the above features, and most locking frames are difficult to maintain and work when there are more than 100 levers interlocked.

7
Signal Box Arrangements

The size of a signal box depends on the following items.

(1) The length of the box is determined by the length of the interlocking frame to be accommodated. It is usual to allow a clearance of about 4 ft. between the locking frame and the end walls of the box, taking the dimensions as inside the box at floor level. This gives the length of the signal box as being 8 ft. longer than the interlocking frame. If there are many large instruments to be fixed in the box it should be built about 10 ft. longer than the locking frame.

(2) The width of the box to some extent depends on the type of interlocking frame used, but the general standard width is about 12 ft. If the box has to be built in a very confined space, the width might be reduced to 9 ft. at floor level.

(3) The height of the box from floor to rail level depends entirely on the height necessary to give the signalman a good view of all his signals and points, and of trains standing at signals. If there are no obstructions to the view from the signal box a suitable height for a signal box is 8 ft. 6 in. from rail to floor level. Sometimes it is necessary to build a high box to enable the signalman to obtain a good view of his work, but, as a general rule, it is better to keep the box low down, as this gives a better view of the tail lamps of trains, especially in "thick" weather, and makes it easier for the signalman to give orders to the train-men and shunters. The height should not, however, be reduced below 8 ft. 6 in. if the signalman is to be expected to observe the tail lamp of a train with another standing or running on an intervening line. The height from ceiling to floor level should not be less than 9 ft. 6 in. to provide sufficient air space.

Signal Box Siting

A signal box must not be built nearer to a running line than 5 ft. to clear the structure gauge, and it should as a rule be kept about 8 ft. back from the line, to give ample space for the rods and wires being led away from the box. If the box is built on a platform it should be built not less than about 10 ft. back from the edge of the platform, so as to give a good clearance for barrow traffic, etc.

With the foregoing limitations the box should be built as close to the running lines as possible, to facilitate verbal instructions from the signalman to the train-men, etc.

Where a signal box requires to be built in a very narrow space, the lower part can be made narrow to suit, but the box should be widened out at floor level to the standard size; under these conditions, if the box is adjacent to a running line or siding, care must be taken that the over-hanging portion of the box clears the structure gauge for height. Sometimes it is necessary to build the signal box on a bridge over the lines, in which case the floor of the box should be kept as low down as circumstances will permit.

Brick or Stone Construction

Signal boxes are constructed of any ordinary building materials, for the majority of cases brick or stone is the most suitable, but for temporary work, or where a good foundation cannot be secured, wood is to be preferred. For overhung boxes, or boxes built on bridges, steel framing is generally used.

Taking an ordinary brick or stone box, the foundations are laid about 3 ft. wide and about 4 ft. deep, depending very much on the nature of the ground. The thickness of the walls is generally 14 in. for brickwork. When the walls have been built to about 18 in. below rail level, some arrangement requires to be made for holding timbers to which the leading-off cranks and pulleys will be fastened.

These timbers are generally of 12 in. by 6 in. section, and are laid on their flat side. A very common way of holding these timbers is to build into the front and back walls cross timbers of similar section to the above laid on edge. These timbers project about 5 ft. in front of the box, and are spaced about 5 ft. 6 in. apart. The leading-off timbers both inside and outside the box are spiked or bolted down to them. The top of the leading-off timbers should be about 11 in. below rail level. There is some objection to building timber into brickwork, as timber requires renewal before the brickwork, and cast-iron girders or old rails are often used instead of the cross timbers.

When cast-iron girders are employed, they are not made long enough to be built into the back wall of the box, but a subsidiary wall is built inside the box, about 5 ft. 6 in. back from the front wall, to

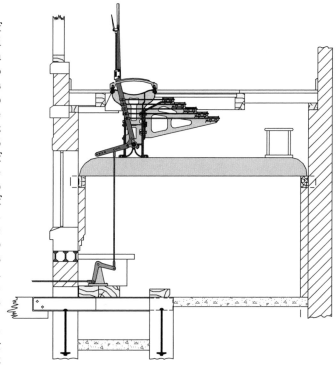

Figure 182
Part section of signal box to suit N. E. R pattern locking frame.

support the inner end of the girder. The other end of the girder extends about 12 in. in front of the box, and bearing timbers are bolted to these to carry the outside leading-off timbers. With this arrangement all the leading-off timbers can be renewed without disturbing the brickwork (see Fig. 182).

The space between the front wall and the subsidiary wall forms a pit for the convenience of fitters attending to the cranks, etc. The bottom of this pit, together with the remainder of the floor, should be concreted and well drained.

Along the front of the box an opening must be left to enable the rods and wires to be led out; this opening should extend the whole length of the box, and should commence about 11 in. below rail and extend to about rail level. To support the face wall of the box, either a wood beam or, preferably, a steel joist lintel (old rails are often used) is built into the box, and is supported at intervals of about 5 ft. 6 in. by cast-iron supports. Small openings should also be left at each end of the box near the front wall to admit of wires or rods being run out there if required. Old rails are generally used to support the wall over these openings.

A door should be provided to give access to the bottom floor. Windows should be made in the front wall to give as much light as possible for examination of the connections and locking apparatus. Sometimes the window frames are of cast iron made in the form of an arch.

FIXING AND SUPPORTING THE LOCKING FRAME

ARRANGEMENTS also have to be made for supporting the locking frame, the type of locking frame employed determines to a large extent the nature of the supports. Taking a type of frame as illustrated in Fig. 165, cast-iron cross girders, resting on brickwork piers below each supporting standard, make a very efficient arrangement (see Fig. 182). The position of the locking frame in the box is fixed by the clearance required by the tails of the levers and the interlocking plungers. If there is any interlocking between the levers and the front wall of the box the locking frame should be placed sufficiently far back in the box to enable workmen to get to the locking for cleaning and repairs, but—bearing this in mind—the locking frame should be placed as near the front of the box as possible to give the signalman a good view when working the levers. It is sometimes advantageous to place the locking frame so that the signalman when working the levers has his back to the line, and in some cases this gives him a much better view of his work, but most boxes are arranged so that the signalman faces his work when pulling the levers.

The floor is built about level with the floor plates of the locking frame; where the floor extends over the locking it is made removable, being supported on T irons bolted to the angles of the locking frame at one end, and let into a supporting beam at the other end.

A "false" floor or scaffold should be provided below the locking frame for the use of fitters when attending to the interlocking.

Another way of supporting locking frames of this class is to use timber cross beams instead of cast-iron cross girders, the other arrangements being similar. A very common arrangement is to support the locking frame on longitudinal timbers which are bolted to vertical posts, the latter being fastened to cross timbers bedded in the ground floor of the box.

To support a locking frame of the type shown in Fig. 162, a beam 12 in. by 12 in. is built into the end walls, and to this beam the supporting standards and soleplates are fastened (see Fig. 183). As this frame is not designed to stand without lateral support, a beam about 9 in. by 6 in. is placed at floor level to take the lateral thrust, and this beam is also employed to carry the joists for supporting the floor. The floor between the locking frame and the front wall is detachable, being in

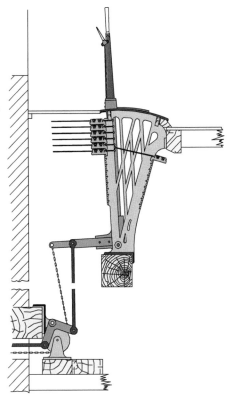

Figure 183
Part section of signal box to suit Stevens & Sons' pattern locking frame.

the form of hatching, and is supported on irons fastened to the locking frame at one end, and rests on the front wall at the other end; this enables the locking-fitters to get to the locking when required by lifting the hatchings.

The brickwork is carried up to about 3 ft. above floor level all round, leaving space for a door. Windows should be put in at this height round at least three sides of the box (Fig. 184), and if there are running lines on both sides of the box then all four sides should have windows. The windows at each end of the box and some of the windows facing the rails at the end of the locking frame should be made to slide along, to allow the signalman to converse with the men outdoors.

If the locking frame is a very long one, a break or gap should be made in the middle to allow the signalman to get to the window, the windows opposite the gap being made to slide.

WOOD CONSTRUCTION

IF a box is constructed of wood, the foundations may be brickwork, with a timber framing laid thereon, or brickwork may be entirely dispensed with, the foundations consisting of beams of wood or piles, should the ground be very soft. The opening in front of the box (for the rods and wires) will not require a heavy girder to support the superstructure, but a heavy beam may be inserted for the purpose of having cranks fixed on to it. If the cranks are not fixed to this beam, an 8 in. by 6 in. beam is ample size for the framing support (Fig. 185).

Cross timbers will require to be set in the ground for the leading-off crank timbers, and vertical timbers will be placed on these cross timbers for the purpose of carrying the locking frame. The windows in the basement will be made of wood framing instead of cast iron, as described for a brick box, and be of a simple square design.

Some brickwork is necessary to support the fireplace and form a chimney, etc.

(Continued on page 117)

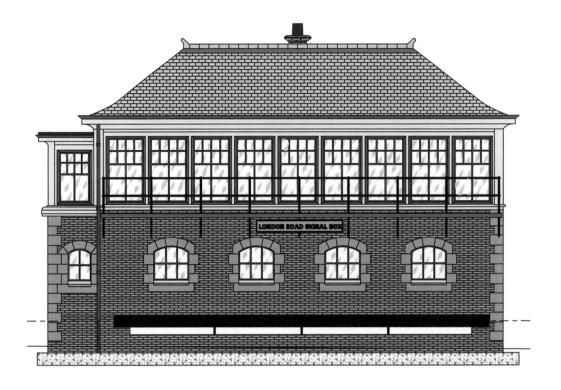

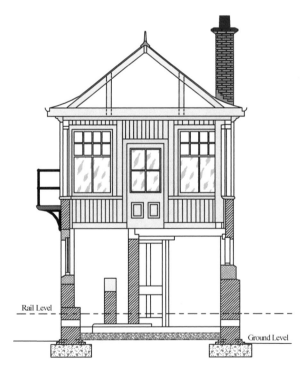

Figure 184
Brick signal box

Figure 185
Wood signal box

(Continued from page 115)

The top portion of the box is practically the same as for a brick-built one.

The door of a signal box should open on to a landing, and the steps up to this landing should be arranged so that a man coming down them will have a good view of approaching trains. If the steps lead directly on to the running lines, a rail or barrier should be placed at the foot of the steps to prevent a man from inadvertently stepping in front of a train.

Signal boxes should be provided with water and lavatory accommodation, the lavatory being often placed at the top of the steps on the landing.

INTERIOR FEATURES

A fireplace or some other heating apparatus must be provided, not only for the comfort of the signalman, but to keep the box dry, as dampness is liable to affect the electrical instruments injuriously, besides causing the interlocking apparatus to become rusty.

Inside the box a shelf should be fixed to support the electrical instruments. The block instrument shelf is generally fixed above the interlocking frame, and this allows the signalman to do the necessary belling, etc., and work the levers without loss of time. The shelf is either suspended from the roof or else supported on cast-iron uprights, which are bolted to the irons carrying the hatching at the back of the locking frame as in Fig. 186. Occasionally these uprights are bolted directly to the floor plates of the interlocking frame.

Electric repeaters for the signals should be fixed as near the levers working the signals concerned as possible, and a board running along at the back of the levers is often used to support these instruments.

A board giving the reading of the levers, called a "direction" or "name" board, is also fastened immediately at the back of the levers.

The following furnishings also require to be provided: Chairs, desk for train register book, etc., clock, and lockers or cupboards for storing cleaning utensils, etc., etc. Cupboards also have to be provided for the electric batteries. These cupboards are generally fixed on the bottom floor of the box; in a few cases they are fixed on the same floor as the levers. A board is very often fixed in the signal box for the signalman to pin weekly and other notices requiring his attention; also a small rack for flags and another for emergency working forms are frequently provided.

In erecting the locking frame on timbers (see Fig. 183), it is necessary to set the beam high in the centre, as when the load of the locking frame comes on it, a slight sag invariably takes place, and if the locking is to work easily it is essential that the frame shall be perfectly level when finished.

The rods from the lever tails to leading-off cranks are generally of solid metal, unless the box is a very high one, when ordinary point rodding is used and guides inserted to prevent the rods from buckling.

WIRE REGULATORS

IN connecting the signal wires, chain is generally used round pulley wheels, and the connection to the lever is by means of a hook

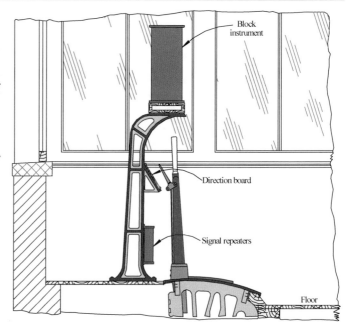

Figure 186
Standard for supporting block shelf, direction board and signal repeaters

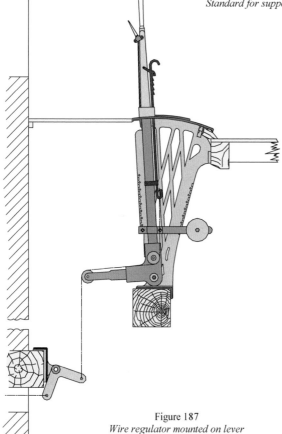

Figure 187
Wire regulator mounted on lever

(unless wire-adjusting screws are used), to allow of the wire being lengthened or taken up as may be required. With point rods, as already described, compensators are used outside to allow for variations of temperature; but with signal wire automatic compensators are very seldom used, and no compensating device is used unless the signal is some distance from the box, in which case wire regulators are fitted in a convenient position so that the signalman can regulate the length of the wires as required.

Some railway companies fit regulators on each signal lever (Fig. 187) when the signal is not more than 300 yards away. Above that distance regulators fixed on the front wall (Fig. 188, A) or on the floor (Fig. 188, B) are employed, as a greater variation in length can be allowed with either of these patterns.

Automatic wire regulators are employed by some companies to a small extent, most of the types tried depending on the principle that the signal wire, instead of being fixed directly to the signal lever, is connected to a counterbalance weight of some description. Normally the weight keeps the wire tight, and any variation in the length of the wire, either expansion or contraction, is accommodated by the weight either falling or rising as the case may be.

When the signal lever is pulled it engages with the signal wire or some mechanism connected to the signal wire, and so pulls the signal off in the usual manner. While the lever is pulled the wire regulator is out of use, but on returning the lever to the normal position again the wire is disengaged from the lever, and the weight again keeps the wire tight.

A wire regulator of this pattern is of no value in keeping the wires adjusted where the signal box is switched out and all the running signals pulled to clear. Where the signal box is open all the time, however, it saves the signalman from having to adjust his wires during the day or night.

When the tail of a signal lever is not

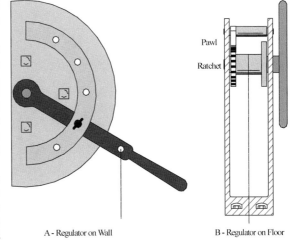

A - Regulator on Wall   B - Regulator on Floor

Figure 188
Wire regulators, floor and wall types

long enough to give sufficient draught to the wire for working a signal at a great distance from the box, additional pulley wheels are added, termed "draught" or "multiplying" wheels, to give the required travel of wire (see Fig. 189).

Where a signal is more than 600 yards from the box, unless there is a very straight run of wire, back balance weights are necessary on the levers to assist the signalman in pulling the lever over; these weights are either fitted on to a back tail or are held on a rod which is connected to the lever by means of chain and pulley wheels, the weights being placed at the back of the box (see Fig. 189).

When signal boxes are placed on bridges or in confined situations, special arrangements have to be made for leading out the rods and wires to suit the particular case.

Paragraph 7 of the Ministry of Transport Requirements refers to signal boxes, and reads as follows:

SIGNAL BOXES AND INTERLOCKING

7. The levers working points and signals to be brought close together in a signal box, or on a properly constructed stage. The signal box to be sufficiently commodious to allow the signalman to have free access where necessary to windows. It

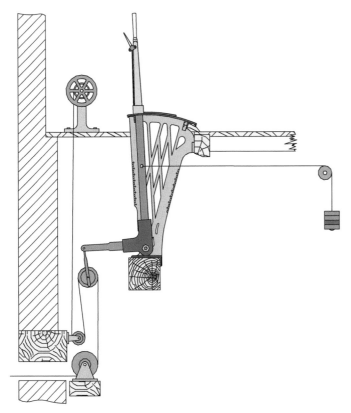

Figure 189
Lever with "draught" wheel and back balance weight

should be provided with a clock, and with up and down three-position block instruments for signalling trains on each line of rails. The point levers and signal levers to be so placed in the box that the signalman, when working them, shall have a thoroughly good view of the railway; and the box itself to be so situated, elevated and constructed as to enable the signalman to get the best possible view of all the operations for which he is responsible. Lights in the signal-box to be so arranged as not to be mistakable for fixed signals. Telephone communication between signal boxes is desirable.

Adequate arrangements to be made where necessary for reminding the signalman of vehicles which are standing within his control. In the case of passenger lines with high speed traffic, or where light engine, crossing, &c, movements are frequent, these arrangements should preferably be automatic when Stop signals are at a considerable distance from the box, or the signalman's view is likely to be obstructed.

8

LEVEL CROSSING GATES AND SWING BRIDGES

WHEN a public road crosses a railway on the level near a station, the gates are as a general rule worked from the station signal box. If the level crossing is some distance away from a station a signal box may be built, should it be also useful as a block box for shortening the block section. If a block box is not required and the road is not very busy, a gateman's hut may be erected and the gates worked by the gateman; and in such a case it is not necessary that the gates be worked by gearing, but they should be protected by home and distant signals in each direction, and interlocked with the protecting signals.

LEVEL CROSSING GATE GEAR

THE interlocking is effected by means of a LOCK lever in a small locking frame, and from this lock lever (which locks all conflicting signals in the usual manner) a rod is taken to a special lock box fixed in the roadway below the free end of the gates. A bolt is fitted on the free end of each gate, and this bolt extends down into the lock box when the gates are shut against the highway. In the pattern illustrated in Fig. 190 the bolts would then be in the lock, and they cannot be withdrawn until they have been turned through 90 degrees. The turning of the bolts moves a locking disc inside the lock casting, and this can only be done when the lock lever is pulled over. When the lock lever is normal, the lock disc engages with the slide operated by the rod from the lock lever and so prevents the bolts on the end of the gates from being turned. When the lock lever is pulled to release the gates, the bolts are withdrawn and the lock disc fouls the notch in the slide, and this prevents the lock lever from being put back.

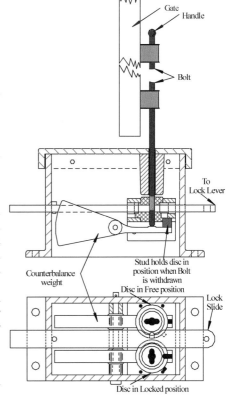

To restore the lock lever in order to lower the signals for the railway, the gates must be placed across the highway, the bolts inserted in the lock box and turned; this takes out the lock disc from the slide and allows the lock lever to be put back. When the lock lever has been put back the gate bolts are again prevented from being turned, and so cannot be lifted clear for swinging the gates. Where four gates are used, each lock box is fitted up to receive two gate bolts, and two such lock boxes are required for the crossing; four may be fitted, but "Dummy" lock boxes are sufficient for holding the gates across the railway.

GEARED CROSSING GATES

WHERE the gates are worked from the signal box by means of gearing, the most convenient arrangement is to fix four gates and work them simultaneously. The gates are hung to posts, so that a crank can be fitted to

Figure 190
Lock for bolt of hand worked gates

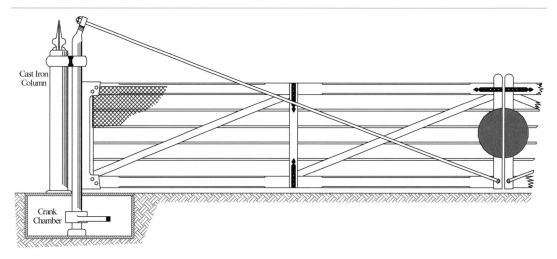

Figure 191
Column and bearings for level crossing gates

the lower portion of their bearing, below road level (see Fig. 191). Rodding is coupled up to the cranks, the rodding being run to the signal box and operated by means of gearing.

If the gates move through a much greater angle than 90 degrees, a crank on the gate bearings is not suitable; instead of a crank a pinion is fitted; this pinion engages with a rack, and the rack is connected to the rodding.

The rodding is of a stronger section than the rodding commonly used for points, being either 1 $\frac{1}{2}$ in. solid rod or 2 in. piping. In most cases it is found convenient to run the rods in the 6-ft. way, but in some cases it may be found desirable to run the rods on the outside of the lines. In any case, the rods must be below road level, and either run in cast-iron trunking or else well boxed in with timber or concrete. It is essential that the cranks and joints shall be capable of inspection without tearing up the roadway, and if timber is used the covering should be fastened down in a manner which will allow of its being opened out without much inconvenience. With cast iron trunking, the lids of the trunking must be chequered to give a good grip for horses' feet, but there should not be any set screws or bolt heads projecting at road level, as these are liable to be torn off by heavy road tractors.

Very often the level crossing is timbered to form a good roadway †; but should it be in a town, and granite setts or wood blocks laid down, it is preferable to have cast-iron trunkings with inspection covers over all cranks.

Where the rods are laid in the 6-ft. way it is a very good arrangement to employ double rods as shown in Fig. 192. This makes a very rigid connection as there is always one rod in tension.

GATE STOPS

WHILE the rodding is sufficient to operate the gates, it is not rigid enough to hold the gates in position, and for this purpose GATE STOPS are fixed. The simplest form is a block of wood fixed in the 6-ft. way, and against this block the ends of the gates strike when they are placed across the railway. The block of wood simply acts as a buffer to prevent the gates from being carried too far round, as their momentum is sufficient to carry them some distance after the signalman has finished turning the winding handle, and this would strain the rodding. This device then only acts as a buffer, and does not prevent the gates from rebounding and standing apart sufficient to allow animals to get on to the line; even at its best it does not prevent persons from pulling the gates away from the stop and passing on to the line. It is not possible to adopt this simple arrangement as a stop for the gates when they are across the highway, as the block being above road level would form a permanent obstruction.

---

† Macadam, or fine ballast, is frequently laid over a level crossing, the timbers covering the rods having about 3 in. to 4 in. of stone or ballast over them.

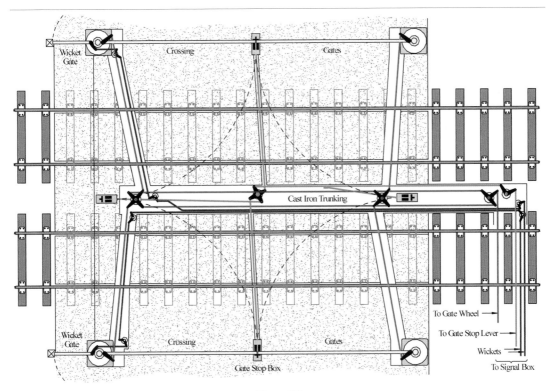

Figure 192
Arrangements of rods, cranks, etc., for level crossing gates

To get over these difficulties, some form of gate stop operated from the signal box is adopted, and the simplest form is as shown in Fig. 193. The stops are enclosed in a cast-iron box and consist of two portions, a FRONT STOP and a BACK STOP. The BACK STOP takes the place of the wood block described above, and it rises above the roadway when the rod working the stops is pushed in, but is lowered level with the roadway when the rod is pulled out. (It should be noted that one BACK STOP serves for both gates.)

The FRONT STOPS are independent, one for each gate, and are counterbalanced, so that they move upwards when permitted by the stud marked "A", Fig. 193. When the back stop is down, the front stops are also down; when, however, the operating rod is pushed, the back stop is forced up, and the front stops are allowed to rise as the counterbalanced end moves down on stud "A". When the gates are moved towards the stops the front stops are forced level with the roadway by the ends of the gates passing over them, but as soon as the gates are past them the front stops instantly fly up and prevent the gates from swinging back after they strike the back stop. It will be noticed that the form of the front stops allows the gates to depress them only when the gates

are being moved towards the stops. Before the gates can be moved back again to their original position, the front stops must be placed level with the roadway, this is done by pulling the rod, the back stop moving down at the same time. Two front stops are provided, because in service it generally happens that one gate arrives at the stops before the other, and with two separate stops each gate is trapped independently; if this were not done, it would often happen that the gate arriving last would release the gate which had arrived before it.

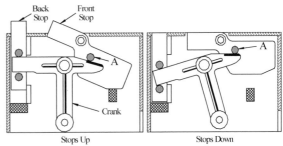

Figure 193
Gate stop box (section)

Fig. 192 shows the complete rodding for working gates and gate stops where four sets of stops are fitted. All the stops in this case are worked by one lever, and when one set of stops is raised ready to receive the gates the opposite set of stops is level with the railway.

With gate stops of this description it is a simple matter for anyone to depress the front stops with the foot, force the gates open (straining the connections), and get through on to the railway. This can be avoided by putting an iron sheath to cover the front stops, one sheath being fixed to the bottom of each gate.

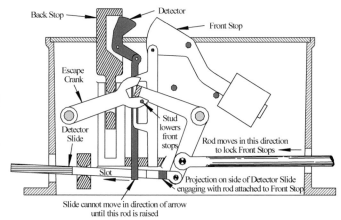

Figure 194
Detected gate stops

Fig. 194 shows an improved pattern gate stop, where the front stops are locked after the gates have been "trapped". All four stops are worked by one lever, but instead of the lever being right over when the gates are being moved open, the lever is placed at a point about four-fifths of its full travel. This places the stops in the correct position to receive the gates. After the gates have been fully opened, the lever is moved the remaining one-fifth of its travel, and this last movement locks the front stops by means of the slide attached to the crank engaging with the link fixed to the front stop. Owing to the escape crank this final one-fifth motion does not impart any further travel to the stops.

To return the gates to their normal position, the gate stop lever is first put to a position about four-fifths back; this lowers the front and back stops which hold the gates, and at the same time puts the opposite set of stops in position to receive the gates. When the gates have been moved and are caught in the stops the last one-fifth of the travel of the gate stop lever locks them in that position.

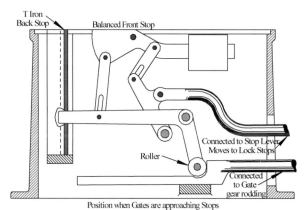

Figure 195
Balanced gate stops

With many pattern stops there is a possibility of the signalman locking the stops before the gates are properly secured; to prevent this the stops can be fitted with a detector, against which the gates strike when they have passed over the front stops. In pressing back the detector (shown grey in Fig. 194) the rod connected to it is lifted clear of the slide which is attached to the main crank, allowing it to be moved to lock the stops. Owing to the complicated nature of "detecting" stops they are seldom used.

A more serious source of trouble occurs when the signalman places the gate stops in the roadway ready to receive the gates some little time before he moves the gates across the roadway; the stops being *up* form an obstruction in the highway, and serious accidents have occurred owing to horses stumbling over the raised stops. This difficulty has been overcome by operating the back stops directly from the main gate gearing, so that they lie in the down position until the gates are within a yard or two of them when they rise to receive the gates. If balanced front stops are used, these are allowed to rise with the back stops (see Fig. 195). In some types, however, non-balanced front stops are used, and these are mechanically prevented from being raised until the back stops are up (see Fig. 196). To replace the gates the front stops have first to be lowered by the gate stop lever, and as the gates move away from the stops the back stop gradually returns to its normal position level with the roadway. Fig. 197 shows a lay-out for automatic stops of this description with gates on the skew.

It is possible to have gate stops incorporating all the safeguards above described, but in service it is found that complicated devices give far more trouble than devices which are simpler, but which may be less perfect theoretically.

As gate stops are fixed in the middle of a roadway, it is obvious that the less attention and examination they require the better. They are liable to be silted up with mud, and in bad weather flooded with surface drainage; and for any machine to work satisfactorily under these severe conditions requires the simplest possible mechanical motions, strong bearings and parts, and, if possible, some means for removing silt.

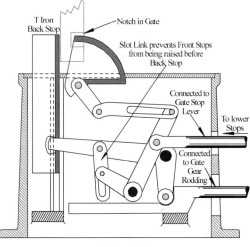

Figure 196
Non-balanced gate stops

Efficient drainage is necessary, or else a cover which can be removed for baling out water and mud must be provided.

WICKET GATES

IN addition to the main gates, small gates are fixed at the side to allow pedestrians to pass when the main gates are closed against the roadway. These gates (termed "wicket gates") are *controlled* from the signal box. Sometimes where there are short gates on both sides of the crossing, all four gates are worked in a manner similar to the main gates, but in most cases the gates close themselves by means of a gravity counterbalance, and they are either bolted in the closed position, or else held in the closed position by means of a chain or slotted rod. A common way of causing the gates to be self-closing is to design the bearing so that when opened the gate rises

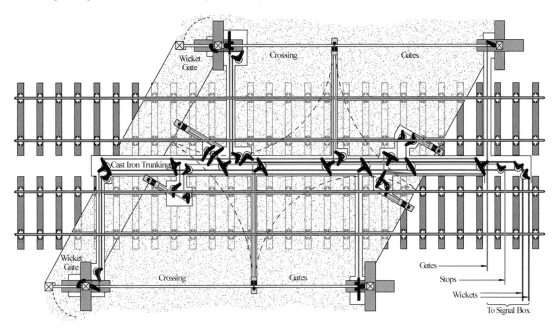

Figure 197
Arrangements of rods, cranks, etc., for gates on the skew with automatic gate stops

slightly on a helical path; the whole weight of the gate then tends to make it close itself. Occasionally a weight in a pit with a chain and quadrant portion of a wheel is fitted to the gates below street level to make them close. It is preferable to hold the gates closed by means of a slot, or chain, rather than a bolt, as should the wickets be slightly open it is impossible to bolt them. In the case of the chain or slot, when the wicket lever is pulled, the gates are held tightly against the gate post, it being of no consequence should they not be closing properly by their own weight (or a counterbalance weight). It is desirable also to have a separate control lever for each wicket as without this there is a risk, when the signalman releases one wicket to enable a pedestrian to go out, of others entering by the second.

GEAR FOR WORKING GATES FROM SIGNAL BOX

THERE are many types of gearing for operating the level crossing gates. A very common arrangement is that of a wheel geared to a leading screw; a large nut on this screw is connected by means of a link to a down rod and thus on to a leading-off crank, as shown in Fig. 198. Another way of utilising a leading screw is shown in Fig. 199, the nut engaging directly on to a crank. Fig. 200 shows a device consisting of a train of wheels engaging with a segment of a circle fitted to a lever, the lever being made to suit the locking frame. The lever is made of a stronger section than the ordinary point or signal levers.

The screw gear types are usually fixed near the window at the end of the box adjoining the gates, and are independent of the locking frame. The tooth gearing types are usually fixed as a part of the interlocking frame, and are placed at that end of the frame which is the nearer to the gates.

Figure 198
Gate gearing—vertical screw pattern

INTERLOCKING BETWEEN THE GATE GEAR AND THE SIGNALS

TO effect the interlocking between the gate gear and the signals an independent lock lever is sometimes introduced, the lock itself generally being an ordinary facing point lock shoe fitted on timbers below the floor. A rod is taken from the leading-off gate crank to the lock, and this rod operates a slide with one hole in it, the hole being made to fit the lock plunger. The lock plunger is connected to the lock lever, and it is arranged that when the lock lever is normal the gates are bolted across the highway. The main line running signal levers lock this lever, so that should the running signals for a train approaching the gates be *off*, it is not possible for the gates to be moved across the railway. To place the gates across the railway the lock lever is pulled; this withdraws the bolt, and when the gates have been placed across the railway it is impossible for the lock lever to be returned to the NORMAL position, as the slide in the lock shoe operated by the gate leading-off crank prevents the lock plunger from entering. Not until the gates have been placed across the roadway can the lock lever be replaced. This is a very simple device, but it allows the signalman to strain the connections between the gate wheel and the lock shoe should he attempt to move the gates with the lock lever normal. A more efficient method is to have a small pawl on the gate gearing which engages in a notch cut in the boss of the gate wheel (see Fig. 200). This pawl may be worked from the gate stop lever if gate stops of suitable pattern are used, such as

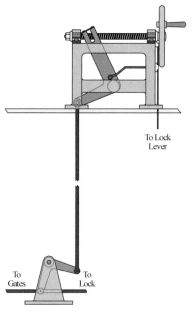

Figure 199
Gate gearing—horizontal screw pattern

shown in Fig. 194; but if gate stops of the pattern shown in Fig. 193 are used an independent lock lever must be employed.

The arrangement of working stops, as in Fig. 194, and gate gear, as in Fig. 200, is thus:

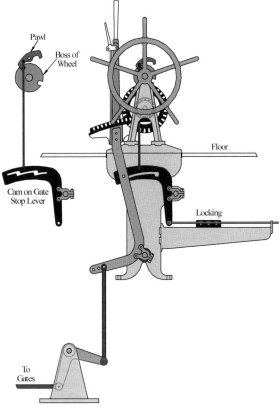

Figure 200
Gate gearing—lever pattern

- Normally the gates are across the highway, and the gate stop lever also is in the "normal" position in the locking frame. This allows traffic on the railway to proceed. To set the gates for the highway the gate stop lever is moved four-fifths over; this takes the pawl out of the notch in the gate wheel boss by means of a cam, and allows the gate wheel to be turned. A description of the movements of the corresponding stops has been given on p. 122

- When the gates are *home*, inside the stops, the gate stop lever is pulled right over; this places the opposite end of the pawl into a notch in the gate wheel boss. To replace the gates a similar procedure takes place, the stop lever being placed four-fifths back to take the pawl out of the notch. This arrangement ensures that the gate stop lever is in the correct position to place the stops ready for receiving the gates; and it is not possible to strain the connections by attempting to wind the gates without first having lowered the stops.

Should the gates first be moved in one direction (with the stops placed correctly to receive them), and the signalman wish to return the gates to the original position, it would be possible for him to do so without having placed the stop lever in the correct position for this movement; to prevent this a ratchet pawl is sometimes fitted so that when once the signalman has commenced to move the gates in one direction he cannot reverse the direction of motion until the gates have been moved to the full distance and the stop lever placed in the correct position for the reverse movement.

The wicket gates are usually worked by ordinary levers in the interlocking frame, but without any interlocking attached.

Sometimes small two-lever dwarf frames are used, these being fixed on the floor of the box near the windows overlooking the gates.

WARNING GONGS AND BELLS

AT very busy crossings some form of warning gong is occasionally provided to warn users of the highway that the gates are about to be closed against the roadway. A very good device is a large-size electric bell fitted up inside a casing with louvre boards; the bell circuit is closed by the first motion of the gate stop lever, and the last motion of the gate stop lever breaks the circuit. A trip contact is employed to prevent the bell from ringing when the gates are being opened again for the roadway.

Paragraphs 28 and 29 of the Ministry of Transport Requirements refer to Level Crossings, and read as follows:

LEVEL CROSSINGS

28. - (1) At all level crossings of public roads, gates, where they are prescribed, must be constructed completely to close alternately across the railway and across the road on each side of the crossing. They must not be hung so as to admit of being opened outwards towards the road. Stops to be provided to keep them in either position.

In all cases where the normal position of gates is across the roadway, arrangements will be required to work them either from a signal box, or by an attendant, for whom special accommodation may be necessary. Where the normal position of gates is across the railway, an attendant will be necessary, unless the gates can be opened when required and closed by the trainmen.

At public road level crossings in or near populous places the gates to be either close-barred or covered with wire netting.

Red discs or targets for daylight and red lamps, one on each side of the crossing, for night, to be fixed on gates; the discs or targets, and lamps, according to the position of the gates, to show toward the road; also towards the railway if there is no fixed Stop signal at the gates. One such disc or target and lamp to be fixed on the gates on each side of the crossing.

Fixed railway signals will not be required when, having regard to the traffic gradients, &c., a sufficiently good view of the discs or lamps is obtainable by enginemen of approaching trains to enable them to stop short of the gates when they are across the railway. When, however, the view obtainable by enginemen is insufficient for this purpose, and it is considered necessary to give additional protection beyond that furnished by the gate discs or lamps, a fixed signal of the Distant signal type to be provided. The Distant signal may be either of the one-position unworkable type, when it is desired to give warning only of the proximity of the level crossing, or of the two position worked type when information in respect of the actual position of the gates at the level crossing is conveyed. In the latter alternative the Distant signal must be interlocked with the gates.

At important level crossings, or where conditions require them, Stop as well as Distant signals interlocked with the gates will be necessary.

At all level crossings of public roads or footpaths a bridge or a subway for pedestrians may be required.

Attendants at level crossings provided with gates should, as a rule, receive warning by bell or otherwise of the approach of trains from either direction.

(2) At public road level crossings when gates are not prescribed and cattle guards may be used to prevent trespass upon the railway, speed reduction and whistle boards will be required at suitable distances on the railway on each side of the level crossing. Warning boards visible both day and night will also be required on each public road approach. It will be necessary to ensure that a good view of the railway line in each direction is obtainable from the road approaches to the level crossing by clearing or lowering obstructions to sight such as hedges, &c.

(3) For field, private and occupation road level crossings, single gates should be used. They should be hung so as to open away from the railway line.

29. Sidings connected with the main lines near a public road level crossing to be so placed that shunting may be carried on with as little interference as possible with the level crossing; and, as a rule, the points of the sidings to be not less than 100 yards from the crossing.

SWING BRIDGES

WHERE swing bridges carry a passenger line over a river or canal they must be fitted with a bolt and electrical locking gear to ensure that the bridge is held safely for the passage of the train, and in the event of the bridge being swung, to ensure that all trains are held at a sufficient distance away for safety.

There are two common methods of carrying out the necessary safeguards. In the first case a block signal box is built by the side of the swing bridge, and the levers with the other control gear for operating the bridge are fixed in the signal box. This arrangement is employed when the railway takes precedence over the waterway. In the second arrangement the controlling gear is installed in a box fixed on the swinging bridge, and outpost boxes not less than a quarter of a mile away are erected. This second arrangement is adopted when the waterway is of greater importance than the railway.

BLOCK SIGNAL BOX ADJACENT TO SWING BRIDGE

TAKING the first arrangement, where the block box is built by the side of the bridge, a modified form of lock and block working (see p. 61) may be installed between the bridge box and the block boxes on either side of it, and in addition a bridge release lever which locks the starting signals is fitted in these adjacent boxes. A similar release lever is fitted in the bridge box, and it is required that ALL the release levers be pulled before the bridge can be swung. With this arrangement, before a train can be accepted by the bridge box the release lever must be normal; this holds the turning and other gear, also the bridge bolt, in the position securing the bridge for the railway traffic (ordinary mechanical interlocking being used). The home and distant signals are fitted with reversers or electric motors, the electric circuits being arranged so that unless the bridge is firmly seated on its four corner blocks (or corner wedges, where such are provided) and the bolt in the correct position holding the bridge, the signals cannot be lowered, and if after the signals have been lowered any movement of the bridge takes place, they go to danger.

When everything is normal the signalman at the bridge box can accept a train from either (or both) of the boxes adjacent to him.

Having accepted a train on the block instruments, his bridge release lever is locked electrically until the accepted train has passed over the bridge and the tail of the train is well clear of it. This is effected by means of a balanced bar and treadle working an electric lock. Sometimes catch-points are fitted on the line ahead of the bridge so that in the event of a train backing it will not foul the bridge.

Unless the bridge bolt is of the balanced latch type, it is necessary to provide against its being placed in the bolted position before the bridge is resting on the corner blocks. This can readily be effected by putting an electric lock on the bolt lever, so that until the bridge is home on its corner blocks the bolt lever is back-locked. Fig. 201 shows diagram of wiring, etc., for the above system.

There are several different ways of arranging the details of the interlocking of the release lever in the bridge box. It is in many cases effected without the aid of release levers in the adjacent signal boxes.

SIGNAL BOX BUILT ON THE BRIDGE

WITH the second method of control, using outpost block boxes, the general principles are the same as the system just described. The bridge box is built over the lines on the bridge and controls all the turning and bridge gear, including the bridge bolt. A release lever interlocks with the bridge-gear levers, and when the release lever is normal the starting signal levers of the outpost boxes are locked, and the bridgeman is at liberty to swing the bridge as required. When a train requires to be accepted the bridgeman pulls his release lever which

(Continued on page 131)

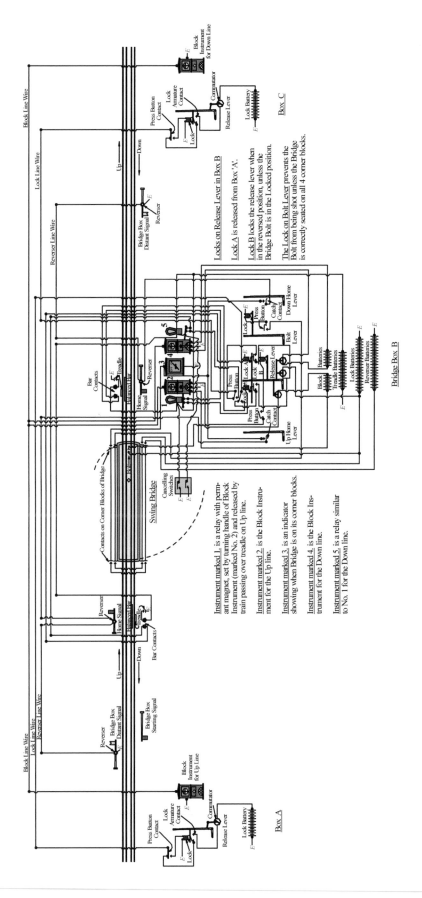

Figure 201

Wiring diagram for swing bridge control

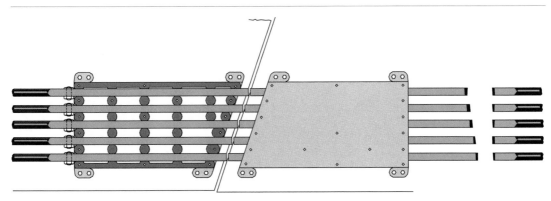

Figure 202
Method of running rods over swing bridge using a rod engager

(Continued from page 129)

then becomes electrically back-locked. When the lock is in the notch of the release lever plunger, the bridgeman is able to release electrically the starting signals of the outpost boxes. The release lever remains back-locked until the trains accepted have passed over treadles clear of the swing bridge, when he can again replace the release lever to its normal position and swing the bridge as required. The starting signals, having been pulled and put back, cannot be pulled a second time until again released from the bridge box.

When the bridge is swung, the circuits for the block instruments of the outpost boxes are broken, so that no attempt can be made to accept trains. The home and distant signals of the two outpost boxes are fitted with reversers or electric motors, and the circuits for these pass through contacts on the four bearing corners of the bridge, so that any displacement of the bridge places the signals to danger. The reversers for these signals must be of the type requiring a current to flow before the signals can be lowered, and any cessation of current must place the signals to danger. Indicators are fitted in the three boxes concerned in circuit with the reversers and bridge contacts, to inform the signalmen when the bridge is home on its seatings.

SIGNAL AND POINT WORKING ACROSS SWING BRIDGE

MOST swing bridges require some gear for allowing signal wires or point rods to be worked across the bridge. Electrical circuits are easily connected up with ordinary rubbing contacts. Fig. 202 shows a simple arrangement for working signal connections over a swing bridge. All the rods from the signal box butt against rods on the bridge, these in turn butting against rods on the land on

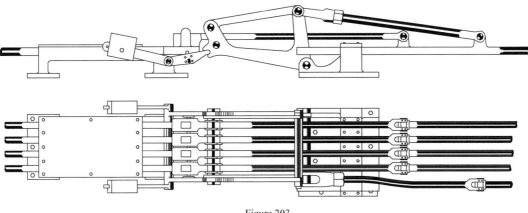

Figure 203
Rod connector for swing bridge

the opposite side of the bridge, and the latter are connected to counterbalanced levers which pull the signal wire. This device can only transmit a PUSH. For the return motion, the rods are moved by counterbalanced levers, so that when the lever in the box is replaced the rods follow and return to their normal position.

For points, two separate rods are necessary; on the land side the rods are connected by cross levers, so that they work in opposite directions; this, then, provides one rod in compression for each direction of motion. This method allows the bridge to be swung when all the rods are in the normal position.

Fig. 203 shows a type of connector for rods passing across swing bridges which requires a special lever for disengaging purposes. This type is for use where the operating box is placed on the bridge itself, as the disengaging lever connections are all on the swinging portion of the gear. This type only requires one rod for the points, and is also suitable for signal connections.

9
GROUND FRAMES

IT is frequently necessary to fix a small dwarf frame to work a pair of points which are further from a signal box than the distance allowed by the Ministry of Transport for mechanical operation of points (350 yards). The lever working such points must be locked, and to unlock the lever, permission must be obtained from the main box. The frame for working the points can be a full-sized one if required, but as a rule no signal box is built, and the frame being fixed at about ground level (called a "ground frame"), a long lever would not be suitable for leading out the connections (Fig. 204).

Either an electric bell (with a code of beats), or preferably a telephone should be installed at the ground frame for communication with the main signal box.

It is usual to have a release lever in the main box and a corresponding release lever in the ground frame, with the two release levers interlocked.

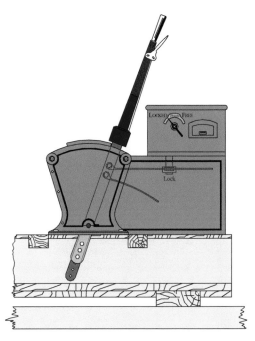

Figure 204
Ground frame with tablet lock for siding on single line

ANNETT'S LOCK AND KEY

THERE are several types of lock for effecting this, the simplest being a lock and key of the "Annett" type (Fig. 205). This is a simple lock bolted to the release levers (one lock on each). Taking the release lever in the main box:

- When the lever is normal, the key is in the lock, the bolt of the lock being in the "unlocked" position. It is arranged that it is not possible for the key to be withdrawn unless the bolt of the lock has been put into the locked position, and the bolt cannot be placed in that position so long as the lever is normal, because the bolt presses against some portion of the locking frame.

- When the release lever has been pulled over, the bolt of the lock comes opposite either a hole in the frame, or a notch cut for the purpose (see Fig. 206); the key can now be turned, placing the bolt in the notch or hole, as the case may be, and on being turned it can be withdrawn and carried to the ground frame.

The Annett lock on the release lever in the ground frame will be locked, with the bolt either in a notch or hole in the frame; on inserting and turning the key the bolt is lifted and the lever can be pulled over, and while the lever is over it is impossible to withdraw the key, as there is no notch or hole for the bolt to enter when this lever is pulled; hence it is not possible for the key to be taken back to the signal box to replace the release lever there until the release lever in the ground frame has been replaced and locked in the normal position. The release lever in the main signal box will lock all conflicting points and signals, and the release lever in the ground frame will unlock the necessary levers for moving the points, etc., there.

Another method of fixing an Annett lock is to drill a hole through the end of the bolt and connect the bolt to an interlocking plunger through the medium of a crank. The interlocking can then be arranged to lock any levers which may be required without the use of a special RELEASE lever.

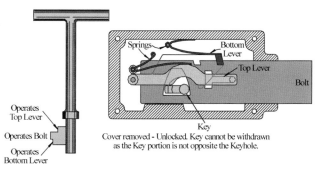

Operates
Top Lever

Operates Bolt

Operates
Bottom Lever

Cover removed - Unlocked. Key cannot be withdrawn
as the Key portion is not opposite the Keyhole.

Figure 205
Annett's key and lock

STAFF LOCK AND KEY

INSTEAD of a simple key, a staff with a key formed on it is sometimes used, and the lock in this case being much larger cannot be bolted to the release lever. The lock is placed at the back of the levers and engages in a locking plunger attached to the release lever (see Fig. 207). The bolt of the lock enters into a hole in the locking plunger, this hole being drilled to suit, according as the lever requires to be locked when "normal", or when "pulled".

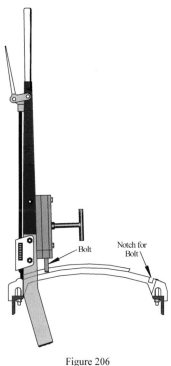

Bolt

Notch for
Bolt

Figure 206
Annett's lock on lever

POINT RODDING LOCK

WHEN it is desired to save the necessity of carrying the key between the main signal box and the ground frame, a lock operated by point rodding is sometimes adopted. A simple device is to use a common facing point lock; this is placed half-way between the signal box and the ground frame. One release lever works the bolt and the other release lever works the slide, it being arranged that the ground frame release lever is locked normally, while the main signal box lever, on being pulled, withdraws the bolt and allows the ground frame lever to be pulled. When this lever has been pulled, it is not possible for the bolt to be replaced; consequently, unless the rodding is forced the main signal box release lever cannot be restored to the normal position. Fig. 208 shows an alternative device instead of a common facing point lock. This device, it will be noticed, like the facing point lock, depends on the point rodding holding good for its faithful working, as, should the rods fail, the locking between the two release levers is destroyed. Since rods are more liable to fail when in compression than tension, the rods are arranged to be in tension all the way, and this, of course, precludes the use of compensators.

A more trustworthy, but more expensive, arrangement is to take a rod from the release lever in the signal box to the release lever in the ground frame and fix a lock directly on the release lever, while for the return locking of the release lever in the main signal box a rod is taken from the release lever in the ground frame to a similar lock on the release lever in the signal box. This lock generally takes the form of a simple plunger working in the ordinary interlocking channels (Fig. 209).

In order to inform the men concerned when the lock has been released an indicator is fixed above the floor and connected to the crank working the lock. The indicator often takes the form of a miniature signal arm, the arm when horizontal indicating LOCKED, and when inclined indicating FREE. Another device is to

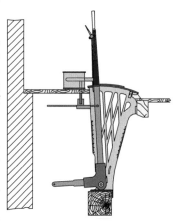

Figure 207
Staff lock on lever

have a small case with a slide inside; this shows LOCKED or FREE through an opening in the face of the case.

Locks of this description are often used for effecting locking between two adjacent signal boxes, and sometimes a similar lock, but operated by wire, is installed for preventing the signalman

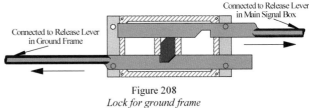

Figure 208
Lock for ground frame

from pulling his distant signal lever until the distant signal of a box ahead has been pulled. Where the lock is operated by wire a counterbalance weight is required to replace the lock (Fig. 210).

ELECTRIC LOCK

THE most efficient, and probably the cheapest, method of controlling a ground frame is by means of electric locks, and this method adds no weight to the levers, as in the case of rod locks. Fig. 211 shows a simple electric lock suitable for this purpose which consists of a common electromagnet; to one end of the armature a pin is attached, and this pin enters into a hole drilled in one of the interlocking plungers, so that in this position the lever is LOCKED. When an electric current flows through the coil the armature is raised, thus lifting the pin out of the hole in the interlocking plunger, and the lever is free to be pulled.

Fig. 212 shows the wiring connections for one lever releasing another. It is absolutely necessary with this type of lock to ensure that the pin is in the hole in the locking plunger of the release lever, before it is possible to send an electric current to

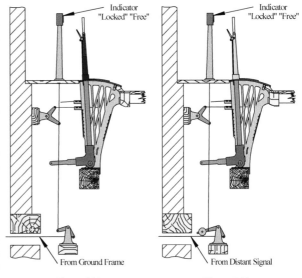

Figure 209
Release lever for ground frame

Figure 210
Wire worked lock for distant signal

free the other release lever. This is effected by including a contact worked by the armature in the releasing circuit. A device called a "commutator " is fixed to each lever, and the commutator only allows the current to flow when the lever is in the correct position. In order to prevent the current from flowing during the whole of the time that the releasing lever is in the correct position, and also to prevent the battery current from releasing the wrong lock, a two-way contact is inserted in the circuit at some convenient position, both in the main signal box and at the ground frame.

This contact is worked off the catch rod of the lever, or it takes the form of a simple push button, which is fixed immediately behind the lever.

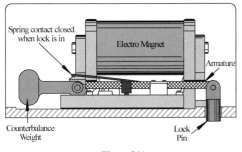

Figure 211
Electric lock

Taking the circuit as shown in Fig. 212 and commencing at the battery marked "A", the wire first goes to the commutator "*a*" fixed on the lever in the box A. This commutator only allows current to flow to the line wire when the lever is pulled. When the lever is over, direct metallic connection is made to the line wire *via* the contact on the lock-armature and press button, and following the line wire to the "ground frame", it leads on to the contact button; when this button is pressed in there is a direct connection made to the electric lock and on to "earth", and as one end of the battery "A" is connected to "earth", an electric current can flow through the lock coils at ground frame B. This lifts the lock and the lever can be pulled over.

When the lever at A is pulled over the pin attached to the lock-armature drops into the hole in the plunger attached to the lever, and this prevents the lever at A from being put back. To put back the lever at A an electric current must be made to pass through the lock coils at A.

No current can be obtained from Battery "A" for this purpose because it is

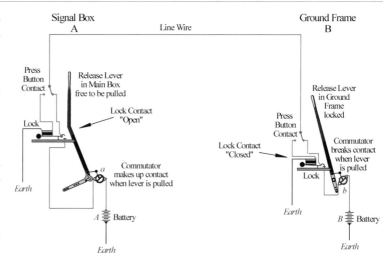

Figure 212
Diagram of wiring for electric locking between signal box and ground frame hut

cut out by the press button. The releasing current can only come from Battery "B", and current can only be obtained from this battery when commutator "*b*", the lock in the locking plunger, and the contact on the press button are in the normal position. Therefore it is necessary first to replace the lever in the ground frame, and as soon as this is done the pin drops into the hole in the plunger. This completes the circuit at B. Then, when the contact button at A is pushed in, the lock pin will lift out of the plunger and allow the lever to be replaced.

The complete circuit therefore is :

MAIN SIGNAL BOX.

Earth—Battery—Commutator—Contact on lock pin—Contact on press button.

Line wire.

GROUND FRAME HUT.

Push button contact—Lock coils—Earth.

SYKES' ELECTRIC LOCK

ANOTHER type of lock often used for this purpose is Messrs. Sykes' Electric Lock. A set of permanent magnets is employed; to the ends of the permanent magnets ordinary electro-magnets are fitted, and an armature is held up by the magnetism induced by the permanent magnets in the cores of the electro-magnets. To release the armature the magnetism induced in the cores of the electro-magnets must be neutralised. This is readily done by sending a current round the coils in such a direction that it sets up a magnetic field of opposite polarity to that induced by the permanent magnets. Fig. 213 shows the general wiring arrangements for one lever to release another lever on this system.

The lock piece engages in the locking plunger when the armature is held up, and to take the lock off the plunger it is necessary to make the armature drop. To re-lock the lever the armature has to be raised. This raising of the armature may be effected by hand, but it is generally done automatically by the lever itself. Taking the arrangement as illustrated, the instrument is fixed above the levers with a down rod connecting the instrument with the lock piece, which is below floor level. The down rod is held up in the locked position by resting on the top of the small crank marked *a* which is attached to the armature. When a current is sent round the coils and the magnetism induced by the permanent magnets neutralised, the crank is forced clear of the down rod

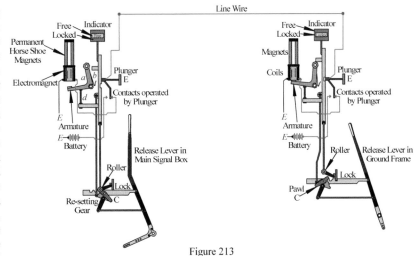

Figure 213
Wiring diagram for electric locking between two boxes using Sykes' Locks

by a small spring marked *b*. The down rod now drops by its own weight and lifts the lock out of the notch in the locking plunger. This allows the lever to be pulled over. On putting the lever back to the normal position a small pawl marked C engages with the roller on the bottom end of the down rod, and this forces the down rod *up*. On the down rod is a stud marked *d* which presses on the bottom of the armature crank and places the armature once more against the cores of the electro-magnet, and, unless the neutralising current is flowing the permanent magnets will cause the armature to be held tightly against the cores. In practice the soft iron of the armature is never allowed to touch magnet cores, and either an air gap is maintained or some non-magnetic material is used to keep them apart.

The wiring arrangements are similar to the system first described, but the releasing current is sent by the signalman pushing in plunger marked E. This plunger can only be pushed in when the down rod is *up*—that is to say, in the position for the lever to be locked; to ensure that the lever is in the correct position for giving the release, a second slide, with a notch in it, is provided; the notch in this slide is opposite the plunger only when the lever is in the required position. This device is in place of the commutators used in Fig. 212.

In order to inform the signalman when the lever is released an indicator is fastened to the down rod; this shows the word "LOCKED" when the armature is held up, and "FREE" when the down rod is in the down position with the lock lifted out of the notch.

This lock is more complicated than the simple coil lock, but has the advantage that the releasing current must be of the correct strength and must flow in the correct direction. If the current is too weak it will not neutralise the induced magnetism of the permanent magnets, and if the current is sent in the wrong direction it only causes the armature to be held the tighter; hence it is almost impossible for a stray current to release the instrument.

LOCKS FOR SIDINGS ON SINGLE LINES

SIDINGS situated on a single line at a distance from a staff or tablet station are fitted with a lock; to gain access to the siding the staff or tablet must be used to release the lock.

If a staff is employed a key is formed on it, and this fits into a lock as shown in Figs. 206 and 207. Where the train tablet is in use this is not possible, but the tablet itself is employed to give the release.

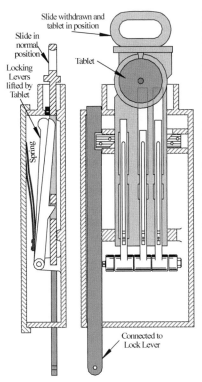

Figure 214
Tyer's patent tablet lock

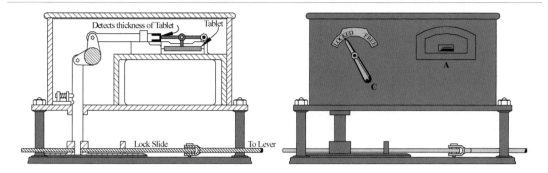

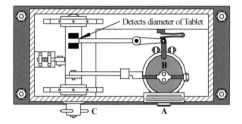

Figure 215
Stevens' patent tablet lock

Tyer's Patent Tablet Lock—Fig. 214 shows a diagram of this lock. Normally the slide is in mid-position, half in and half out of the lock. In this position the lever is locked. When the look is re-quired to be taken off the lever, the slide is drawn out as far as possible, the tablet placed on it and the slide is then pushed in. If the tablet is the correct one for the section it will lift the lock-detecting levers and allow the slide to be pushed right into the lock; in this position the lever can be pulled. When the lever has been pulled, the slide is locked, and cannot be pulled out to regain the tablet. When the lever has been put back to normal position the slide can again be pulled out and the tablet taken away.

Stevens' Patent Tablet Lock—Fig. 215 shows a diagram of Stevens' patent for the same purpose. The tablet is placed in the lock after the handle marked A has been lifted; the placing of the tablet in the lock presses back the piece marked B, and when the handle A is dropped a pin passing through the top of the casting, of which the handle A is a portion, remains up a distance equal to the thickness of the tablet. If the tablet is of the correct diameter and thickness, then the detect-ing piece at the back, and the lever connected to the top detecting pin, allow the releasing handle C to be turned from "LOCKED" to "FREE". This lifts the bolt and allows the lever to be pulled. When releasing handle has been turned to FREE the small handle A cannot be lifted to gain access to the tablet, hence the tablet is locked in the instrument.

When the lever is pulled over it is impossible for the handle C to be put back from FREE to LOCKED, as the bolt worked by that handle presses on the solid metal of the lock slide. To regain possession of the tablet the lever must be replaced in the normal position, and this allows the handle C to be turned to "LOCKED", which again permits the handle A to be lifted and the tablet extracted.

The essential requirements of any pattern tablet lock are:

1. Only the correct tablet must release the lock.
2. The tablet must be locked in the instrument until the lever has been restored to its normal position.

10
Signalling Schemes

Paragraphs Nos. 9, 10, 11, 12, 16, 17, 26 and 27 of the Ministry of Transport Requirements refer to items which must be considered in preparing Signalling Schemes, and read as follows:

Sidings and Safety Points

9. *Sidings to be so arranged that shunting operations upon them shall involve the least possible use of, or obstruction to, running lines. The possible necessity for having in the future to extend passenger platforms should not be lost sight of in designing the layout of stations.*

Safety points to be provided upon goods and mineral lines and sidings, at their junctions with passenger lines, with the points normally set against the passenger lines and interlocked with the signals.

Facing safety points, with or without an overrun or sand drag, may also, in default of other acceptable arrangements, be necessary:

 (a) On single lines, at crossing places, where an adequate interval of space is not provided between the Stop signal controlling the approach to the loop and the fouling point of the loop lines at the other end; or where the line is worked on any non-token system.

 (b) On bay and loop platform lines, as a protection to traffic on the through lines.

 (c) On approaches to opening bridge spans.

Junctions, &c.

10. *Where it is difficult for a signalman to estimate clearance, it may be necessary to provide bars, or other approved device, in order to define the fouling points of junctions, siding connections, crossings, &c.*

Stations

11. *The lines of railway leading to passenger platforms to be arranged so that the platform roads may be entered in the normal direction of movement without reversing; and so that, in the case of double lines, or of passing places for passenger trains on single lines, each line shall have its own platform.*

Curvature of platform lines and of station yards generally to be avoided as far as possible.

At terminal stations, a double line of railway must not, as a rule, end as a single line.

12. *Platforms to be continuous, and of sufficient length to accommodate the longest passenger trains using them. In layouts of passenger stations connections with the platform lines should, as a rule, be clear of the platforms. This does not apply to mid-way scissors or other connections in stations where platforms are long enough to accommodate two trains.*

The minimum clear width of any platform throughout its length to be 6 ft. At important stations the width to be not less than 12 ft., except for short distances at either end in any case of difficulty. In the case of island platforms, the minimum width for an adequate distance on each side of the centre of its length to be 12 ft. The descent at the ends of platforms to be by ramps and not by steps.

Columns for the support of roofs, and other fixed works, not to be less than 6 ft. clear from the edge of platforms. A general clear headway of not less than 8 ft. to be provided over platforms. The height of platforms above rail level may vary according to traffic and other conditions; as a rule, it should be 3 ft.; but, in no case less than 2 ft. 9 ins. or more than 3 ft. at permanent stations, without special approval.

The edges of the platforms to overhang not less than 12 ins., and the recess so formed to be kept clear as far as possible of permanent obstruction. A special recess may be necessary for the accommodation of signal wires, cables, &c. The interval between the edges of platforms and the footboards of carriages to be as small as practicable.

Waiting rooms or shelters, and conveniences, to be provided at junction stations, and elsewhere as may be necessary.

Names of stations to be shown on boards, and on the platform lamps. Platforms must be adequately lighted, and fenced when necessary.

In the case of halts, the above-mentioned requirements in respect of clear width of platform, lighting, and fencing as necessary, will apply; together with such of the other requirements as may be considered necessary in each case.

GRADIENTS

16. *IT is desirable to avoid constructing a station on, or providing a siding in connection with a line which is laid upon a gradient steeper than 1 in 260.*

In the case of steeper gradients, either at a station or siding connection, or within a section between two block posts where the level of one is appreciably lower than that of the other, danger may arise:

(a) *In the case of engines being overpowered by their load;*

(b) *From vehicles running backwards, in the case of trains which have become divided; and*

(c) *From vehicles running away after having been uncoupled on a running line.*

On double lines, one or both of the following arrangements may then become necessary:

(i) *The provision of worked facing safety points, with or without a sand drag, when a special arrangement to utilise an existing facing siding connection is not practicable.*

(ii) *The provision of a single or double self-acting throw-off switch a full train's length in rear of the first Stop signal of the higher block post on the ascending line, when a special arrangement to utilise an existing trailing connection at the lower block post is not practicable.*

On single lines, except where it is possible to work the traffic with an engine at the lower end of an unfitted train, one or more of the under-mentioned measures may be necessary:

(i) *The provision of worked safety points or sand drag, facing the normal direction of descending traffic in a suitable position in the loop at one or both of the block posts, where a suitable arrangement to utilise an existing facing siding connection is not practicable.*

(ii) *The provision of trailing safety points at the lower end of either an existing or specially constructed loop.*

(iii) *The provision of an additional siding, in which the whole of a train can be placed clear of the main line before shunting operations are commenced.*

(iv) *The provision of properly interlocked worked points, a sufficient distance from a siding connection, which can be set as a trap behind vehicles standing below that connection.*

TURNTABLES

17. *TURNTABLES, where no triangle is available, will be required at terminal stations and other necessary places, of sufficient diameter to enable the longest tender engine likely to run on the line to be turned without being uncoupled.*

Exceptions to this requirement will be on lines upon which the traffic is worked solely either by electric traction or by engines suitably fitted and protected for running in both directions, and on short journeys worked by tender engines when speeds are low.

All turntables to be adequately lighted. They should be kept at a safe distance from adjacent lines of rails, otherwise the turntable bolt must be interlocked in the signal lever frame.

STANDARD DIMENSIONS.—CLEARANCES

26. *(1) STRUCTURAL CLEARANCES.—NEW LINES*

In the case of New Lines, after allowance has been made for curvature, super-elevation, and length of rolling stock, the standard static lateral clearance, measured between the point of maximum over-all body width of the broadest stock likely to be used on the line and any standing work, including standards carrying overhead electrical equipment (other than passenger or loading plat-

forms, bridge girders, or disc or miniature signals, up to a height of 3 ft. from rail level), over the whole vertical height between rail level and the top of the highest carriage doors, to be as follows:

    (i)   2 ft. 4 ins. in respect of all structures. Special consideration to be paid to the position of cables, &c., or attachments of any description in tunnels and under bridges.

    (ii) 2 ft. in respect of all signal and lamp-posts, ladders, water columns, &c. In cases of special difficulty, where signal posts, &c., are placed between tracks with only 9 ft. clear intervals (vide para. 27), the minimum clearance of 18 ins. from body-work is permissible.

Minimum lateral clearance of 3 ins. to be provided between load gauge and platform coping, and also between load gauge and structures in the 6-ft. way, below the level of 3 ft. above rail level.

In sidings and at the entrance to all goods sheds, buildings, &c., into which vehicles work, the standard clearance of 2 ft. 4 ins. to be provided above the general level of the ballasting.

In the case of New Lines, having steam traction, or electric traction without overhead equipment on structures, after allowance has been made for curvature and super-elevation, the desirable standard overhead clearance, measured from the maximum load gauge likely to be used on the line, to be 12 ins.

In cases of special difficulty, this dimension may be reduced to a minimum of 6 ins., unless there is any possibility of future electrification on the overhead system, in which case it must not be less than 10 ins.

In the case of New Lines, having electric traction with over-head equipment, after allowance has been made for curvature and super-elevation, the desirable standard clearance, under any conditions likely to arise, to be as follows:

    (i)   Between the underside of any live wire or conductor overhead and the maximum load gauge likely to be used on the line:

        (a)   3 ft. in the open.

        (b)   10 ins. through tunnels and under bridges.

    (ii) Between any part of the structure and the nearest point of any live wire or conductor overhead, 6 ins., after making allowance for any appreciable vertical movement of the live wire or conductor.

In the case of the electrification of existing lines, the last two dimensions may be reduced to 4 ins. as absolute minima.

    (iii) From rail level to overhead conductors at accommodation and public road level crossings, 18 ft.; and 20 ft. above rail level where there is a likelihood of men, in the conduct of their duties, having to stand on the top of engines or vehicles.

    (iv) Between any part of the bow gear and any structure to be 3 ins. as an absolute minimum, after full allowance has been made for lateral movement of the bow.

(2) STRUCTURAL CLEARANCES.—EXISTING RAILWAYS.

    (i)   Where structures are rebuilt, clearances to be provided for as given in para. (1).

    (ii) When permanent alterations are proposed affecting:

        (a) (a) Lateral clearance, by increasing the interval between tracks or by introducing wider rolling stock;

        (b) Overhead clearance, by alteration of the rail level or by the introduction of higher rolling stock;

        (c) Overhead clearance, by the electrification of an existing line;

Unless the minimum dimensions and clearances specified for new lines can be provided or maintained, each case to be referred to the Ministry of Transport, so that it may be dealt with on merits, having regard to the expense involved in providing standard clearances.

(3) CLEARANCE BETWEEN TRAINS.

    (i)   On New Lines, the minimum lateral clearance on running tracks between vehicles of the greatest width likely to be used to be 18 ins., measured at any point over all bodywork. Having regard to super-elevation and the length and contour of rolling stock, the distance between tracks to be suitably increased on curves to provide this clearance at all points.

    (ii) The above also applies to Reconstruction of or alteration to existing lines, where reasonably possible.

    (iii) Where there is a minimum clearance of 18 ins. between stock; door and commode handles, &c., must not project more than 3 ins. beyond the maximum body dimension on each side.

(4) GENERAL.

    (i) Where there are places at which the above-mentioned lateral and overhead clearances do not obtain, the existing clearances must not be reduced, nor must the length or number of such places be increased without approval.

    (ii) The above-mentioned lateral and overhead clearances will not apply in particular cases, e.g., tube railways, where stock of a special type is used and where suitable arrangements are made for the safety of men employed on maintenance and inspection work.

    (iii) In respect of overhead telegraph, telephone, and stay wires crossing the railway, the minimum clearance, in the open, to be 20 ft. above rail level.

INTERVALS BETWEEN LINES

27 *In all new construction, having regard to the lateral clearance conditions set forth above, the interval between adjacent straight lines, where there are two only, not to be less than 6 ft. clear between the rails. If, however, the greatest overbody width of stock likely to be used will not permit of the provision of the clearances named, this clear interval between two adjoining tracks to be increased as necessary.*

Where additional single running lines, or pairs of double lines are laid alongside existing main lines, a standard clear interval between rails of not less than 10 ft. to be provided.

In the case of reconstruction, on existing railways where special difficulties exist, this dimension may be reduced to 9 ft.

In new work, and also in reconstruction on existing railways (except when otherwise approved in cases of special difficulty), the clear interval between a running line and the nearest siding to be not less than 9 ft. Where wagon examination or shunting operations are likely to be regularly performed by men upon sidings, this dimension should be increased to not less than 10 ft.

NOTE.—Appendix II. illustrates the foregoing Requirements in respect of Clearances, &c. It also shows a desirable standard structure gauge which is recommended for adoption for New and Reconstructed Lines (see Fig. 216 opposite).
(See also paragraphs 2, 3, 4 and 5 (p. 26), and paragraph 6 (p. 96).

SIGNALLING PLAN

IN preparing a signalling scheme for any place, it is first necessary to obtain a plan (drawn to scale) of all the lines and point connections. The most convenient scale for such a plan is 33 ft. to 1 in. If the scale is much smaller it is very difficult to show the signals on the plan clearly, and if the scale is much larger the plan becomes inconveniently large.

A plan for signalling purposes should always be drawn so that, when looking at it with the lettering the right way up, the lines, etc., will appear in the same position as the actual rails will look as viewed from the signal box when a person faces the locking frame. At places where it is arranged that the signalman faces the lines when pulling the levers, the plan will have the signal box at the bottom of it, but where the signalman works with his back to the lines, the signal box will appear at the top of the plan.

Fig. 217 (see p. 144) shows a reduced copy of a plan as received from the engineer. All permanent structures, etc., together with mile posts, and the gradients for some distance on either side of the place, and the north point should be indicated on the plan.

The plan shows the centre line of each rail. Thus a single line of railway is indicated by two lines, one for each rail. In preparing sketches for signalling use it is convenient to adopt only one line for each pair of rails, and if the sketch is drawn to a scale, the line would represent the centre of the 4-ft. way. Thus a double line of railway would be indicated by two lines, which would scale about 11 ft. between them, that being the distance between the centres of the 4-ft. ways, where a 6-ft. way is provided between the running lines (see Fig. 223, p 146).

One of the most important items of a signalling scheme is deciding the "set" or "lie" of the points. That is, to determine in which position the point tongues shall lie when the lever working them is in its NORMAL position in the interlocking frame.

(Continued on page 145)

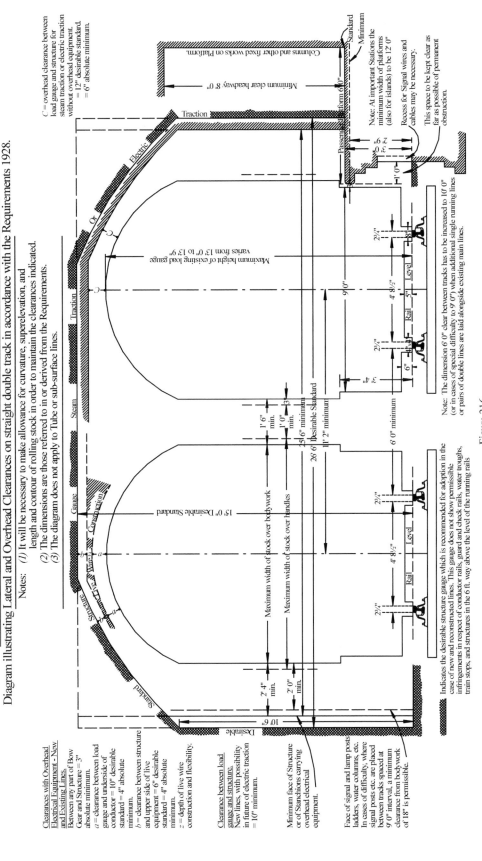

Figure 216

Ministry of Transport clearance gauge

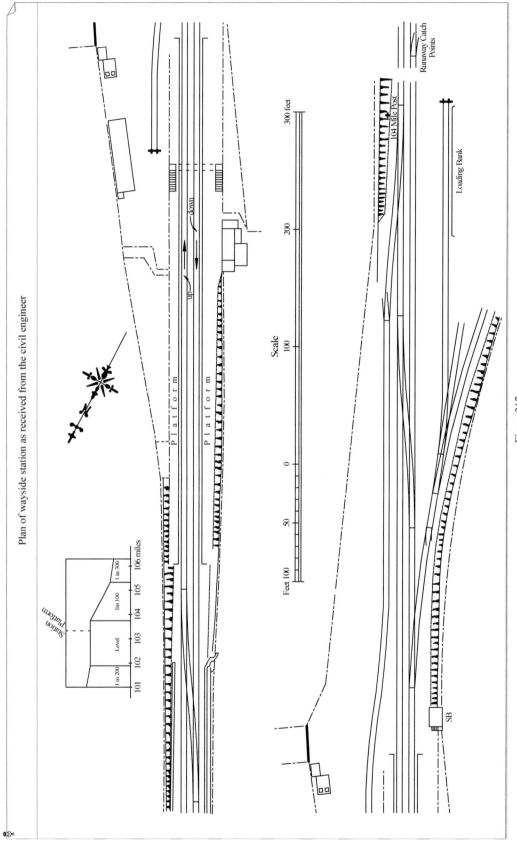

Plan of wayside station as received from the civil engineer

Platform

Platform

down

up

Station Platform

1 in 200   Level   1in 100   1 in 300

101   102   103   104   105   106 miles

Scale

Feet 100   50   0   100   200   300 feet

104 Mile Post

Runaway Catch Points

Loading Bank

SB

Figure 217
Civil engineer's plan

(Continued from page 142)

All the signals, points, etc., are shown on a plan as they stand when the respective levers working them are in the NORMAL position.

On a plan as supplied by the engineer the "lie of the points" is indicated by drawing a line to represent the switch tongue, which lies open (see Fig. 218). Sometimes this line is drawn at an exaggerated angle in order to emphasise the direction in which it lies. With sketches of the single line type the set of the points is indicated by showing the rails over which the train would run as a continuous line, and the portion over which the train does *not* run, when the points are in the normal position, as ending near the continuous line. Fig. 219 shows the lie of points for a simple junction and for a cross-over road.

The signals are shown on the plan as they would be seen by the driver, if they were laid down parallel to the track (see Fig. 220). Slotted signals are shown by the convention previously described (Fig. 19, p. 11). Disc signals are usually shown as a circle, with the addition of a short base (Fig. 221).

Lock bars are indicated by drawing a line parallel to the line representing the rails close against the points to which the bar belongs (Fig. 222).

Detectors are sometimes shown as a small cross near the points detected, but it is generally more convenient to give a table showing the point detection.

Clearance or fouling bars are shown in the same manner as lock bars, the line being drawn in the relative position occupied by the fouling bar. If the bars are worked by levers in the signal box, no remark further than adding the lever number is required; if, however, the bars are depressed by the wheels of the train, and operate electrical contacts, this is indicated by writing electric fouling bar or the initials E. F. B. by the side of the line.

SIGNALLING PLAN FOR A WAYSIDE STATION

FOR roughing out a scheme before finally placing the signals, etc., on the scale plan, it is generally very convenient to make a sketch, foolscap size, not rigidly to scale, but approximately so, longitudinally, and spread out horizontally. Taking an example as in Fig. 223 (see p. 146):

The position of the signal box is the first item to be settled. Several important requirements have to be complied with in fixing this. First the box must be situated so that the points are within the limits allowed by the Ministry of Transport. At the place under consideration the extreme distance between the points scales about 400 yards; the greatest distance allowed for mechanical operation by the Ministry of Transport Requirements is 350 yards. It is therefore obvious that there is a distance of 300
(Continued on page 147)

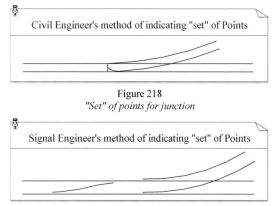

Figure 218
"Set" of points for junction

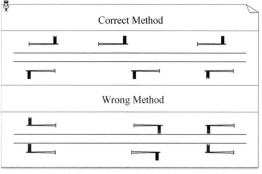

Figure 219
"Set" of points for junction and cross-over

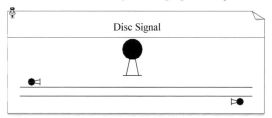

Figure 220
Conventional method of indicating signals on a plan

Figure 221
Conventional method of indicating disc signals on a plan

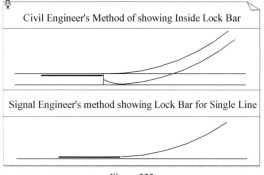

Figure 222
Conventional method of indicating lock bars on a plan

Plan of wayside station as received from the signalling engineer

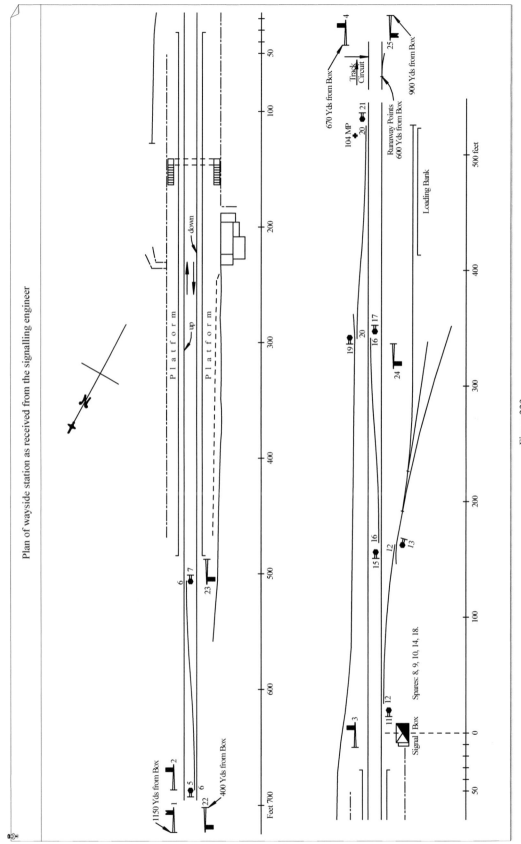

Figure 223
Signalling engineer's plan

146

(Continued from page 145)

yards (150 yards on either side of the mid position) in which the signal box can be placed without having the points beyond the allowed distance from the box.

The next factor to be taken into consideration is that of the view to be obtained from any site within the above limits. The curvature of the line, heights of buildings, trees, etc., have to be considered, and in many cases it is impossible to fix on the best site without visiting the place. The Ministry of Transport Requirements stipulate that the signalman shall have a good view of the railway, and shall be able to see all the point connections, unless an approved alternative for direct vision by the signalman, *e.g.*, track circuited diagram, is provided. Should there be no obstructions to the view of the signalman, the signal box should be placed as near as possible to the majority of the point connections. Not only does this assist the signalman in supervising any shunting movements, but it obviously reduces the cost of the installation, and correspondingly reduces the cost of maintenance.

Having fixed the site for the signal box, the points to be connected to the signal box, together with the way the point tongues are to lie, must next be considered.

All points directly on the passenger lines MUST be connected to the signal box, unless they are siding or junction trailing points, when, in special circumstances, they may be treated like runaway points, and balanced ground levers used.

All points which act as trap or safety points for sidings or goods lines, also main line cross-over roads, MUST under all conditions be connected to an interlocking frame. Points which are on sidings inside the catch, or trap points, may be worked by hand levers of any approved type. As a general rule it is preferable to work all points inside the catch points in this manner. If, for the convenience of traffic, siding points inside the catch points are connected to the signal box, care must be taken to see that the points so connected are adequately protected by ground (or other) signals, if not, there is a liability of the points being "run through" in shunting, or the train split on the points when running over them in the facing direction.

Points not worked from the signal box, as a rule, are indicated on the plan by *not* showing the "set" of the points, but by running the two lines together and making a cross line at the joining point.

In deciding which way the point tongues shall lie when normal, cross-over roads are invariably arranged so that when in the normal position no vehicles can be passed from one line to the other. The siding points joining the main line are usually set so that no vehicles can be put into the sidings from the main line. (There are exceptions to this, as at a place where there is a heavy falling gradient, when it is sometimes convenient to use the trailing points of a siding as runaway catch points.) The points leading from the siding to the main lines must invariably be set so that no vehicles can gain access to the main line when the points are in the NORMAL position.

Taking the UP LINE first, a distant (1) and a home (2) signal is required to protect the north cross-over road (6) (see Ministry of Transport Requirements, p. 26). The home signal must be placed clear of the *fouling point* of the cross-over road, the fouling point being opposite the north end of the cross-over road, as at this point the 6 ft. clearance commences to close in. The signal may be placed exactly at the fouling point, but it is preferable to place it slightly clear of the exact fouling point, about 30 ft. back being a reasonable distance.

It is the general rule to place the signals on the driver's left hand, and about 6 ft. from the running rail, but should there be any obstruction to the view of either the driver or signalman, the signal may be placed on the right-hand side of the line; this is always considered as being on the *wrong* side of the line, and signals are only placed on the wrong side when it is not possible to fix them satisfactorily on the left-hand side.

A common height for a home signal is 20 or 25 ft., but should there be any obstruction of the view of either the driver or signalman, the proper height can only be determined by sighting with poles on the ground. It may be necessary to make it slightly lower, or considerably higher, and signals 50 and 60 ft. in height are in use. Generally, however, it is found that low signals are preferable to high ones.

The position of the distant signal depends on the gradient of the line leading up to the home signal. Referring to the engineer's plan (Fig. 217) it will be seen that the gradient is practically level, and in that case the distance for the distant signal will be about 900 to 1,200 yards out from the home signal depending on the speed and class of trains which have to be catered for. With the distant signal the question of the driver's view is all-important, and the precise position can only be decided on inspecting the place. If possible, the signal should be placed so that it can be seen from the signal box, but under no conditions must the driver's view be impaired to improve the view from the signal box, as, in any case, the arm of the signal must be repeated in the box.

Should there be no obstruction to view, a post giving 20 ft. from rail level to the centre of the arm would be a suitable height for a distant signal.

It is usual to place a signal at the end of a platform to act as a starting (3) signal; where it can be arranged it is advantageous to place the signal slightly ahead of the platform to allow the engine to draw clear of the platform, and leave the whole length of the platform free for the passenger coaches. The distance between this signal and the next point connection (16) joining on to the UP LINE being short, this signal will efficiently protect it, so that it will not be necessary to provide a signal between the starting signal and the cross-over road. A suitable height for this signal is 15 ft.

An advanced starting (4) signal will be required ahead of the siding connection to allow a train to go ahead far enough for the tail of the train to be clear of the siding points, so that it can be backed into the siding if required. The length of trains which usually run on the section of line will rule the distance that this signal is placed ahead of the points. If this brings the signal to a position which gives a bad view from the signal box, the arm, and if necessary, the light, must be repeated. If the signal is more than 400 yards from the box, or if a train or engine standing at it cannot clearly be seen from the box, Track Circuiting to operate a lock on the rear signal will be required.

A suitable height for this signal is 15 ft. to 20 ft. Where the line is straight and all the stop signals can be seen at the same time by a driver as he approaches the station, there should be no possibility of the signal lights coming in line with each other. If the advanced starting signal is 15 ft. high, the starting signal 20 ft. high, and the home signal 25 ft. high, there will be no confusion, but if the signals nearest the driver are lower than any of the signals further away there is a possibility of the lights merging together in one position of the engine when approaching them.

The signals for the DOWN LINE will be similar to those just described on the UP LINE. The home (24) signal must be placed so as to protect the south cross-over (16), it being the first fouling point encountered on the DOWN LINE. Care must be taken here to ensure that there is sufficient space between the main line and the sidings for this signal. The Ministry of Transport Requirements call for at least 2 ft. 0 in. clearance between any signal or lamp, post, ladder, etc., and the widest carriage in use on the line except in cases of special difficulty. The maximum load gauge overhangs the outside edge of the rails 2 ft. 3 in. (see Fig. 216). This load gauge would permit of a signal being erected 4 ft. 3 in. from the edge of the rail. It is, however, usual to allow a greater margin than this, and a space of 11 ft. between the rails is commonly provided, and taking a post as being 14 in. square at ground level, this allows a margin of about 8 in. on each side of the post over the minimum allowed by the Ministry of Transport.

The distant (25) signal for the DOWN LINE should be carried out about 750 yards from the home signal, there being a *rising* gradient of 1 in 100 shown on the civil engineer's plan. As this is steeper than the Ministry of Transport allow for stations without runaway catch points, these must be laid about 450 to 500 yards in the rear of the home signal to suit the length of train running over the line.

The starting (23) and advanced starting (22) signals present no difficulty. Ground disc signals † will be required to control shunting movements through the cross-over roads, and to and from the sidings.

Taking the north cross-over road, a disc (5) signal will be fixed at the extreme end of the cross-over, facing a driver on the DOWN LINE, when standing ahead of the points ready for backing through the cross-over.

There is some divergence of practice as to which side of the line a disc signal should be placed. On some railways this signal would be placed on the outside of the line, the idea being to place it well away from the UP LINE, as it only refers to engines or trains on the DOWN LINE.

---

† Miniature Semaphore Signals can be used instead of Discs, but for convenience the term "disc" is used to mean any type of ground signal.

Should the signal be placed on the outside of the line it must be not less than 5 ft. 6 in. from rail, if post office nets on mail vans require to be cleared.

On several railways, however, the disc signal would be placed in the 6-ft. way between the two running lines, the idea being that the disc should, if possible, be placed on that side of the line to which the driver goes when obeying the signal. In this case of a cross-over road, the driver would go to the left; therefore the disc signal would be fixed on the left-hand side of the line, and this places it in the 6-ft. way.

Another ruling adopted by some companies is to place the disc signal on the left-hand side of drivers, that is, when the driver faces the signal. This would place all signals for backing off the main line, either to a siding or through a cross-over road into the 6-ft. way, and all signals for coming out of a siding to the main line on the outside of the siding. It is seldom, however, that any particular ruling can be strictly adhered to, the question of a good view being of the utmost importance.

A similar disc (7) signal would be placed in the 6-ft. way at the south end of this cross-over road to take drivers from the UP to the DOWN line.

The disc signals should be fitted with detectors, so that unless the points are properly OVER it is not possible for the signal to be cleared. Although the movements are not connected with passenger trains, the Ministry of Transport request that such points be so fitted, and the cost of a detector is insignificant in comparison with the expense involved should a train become derailed in using the cross-over, owing to the points not being correctly in position.

The points (12) leading to the DOWN LINE SIDING are situated near the signal box, so that a signal might be considered as being unnecessary for backing into the siding, because it would be quite easy for the signalman to hand-signal the train in. It is, however, desirable to limit as much as possible the amount of hand signalling necessary even at a small signal box, as should there happen to be two trains at the station at the same time there is a possibility of the hand signals being misunderstood. Apart from this there is greater safety in fixing a signal, as while the signal is pulled clear it is impossible for the points to be moved. Therefore so long as the driver or shunter can see the signal at clear he knows that he can move the train with confidence; without the signal, however, the outdoor men have no guarantee against the signalman moving the points and splitting the train without warning. There is also the security afforded by the fitting of the detector in connection with the disc signal.

At a station where the sidings are very seldom used, it might not be economical to fix signals for backing in, and should there be any doubt as to the points holding closely, this can be ensured by clamping the points before the train is backed in. At all places, however, where the sidings are used every day it pays in the long run to equip all the siding and cross-over points with signals and detectors.

The disc (11) signal for backing into the siding at this place would be placed by the side of the DOWN LINE at the end of the points, and on the right-hand side of the drivers when backing into the siding. A signal (13) for moving out of the siding on to the main line must be fitted at the catch points of the siding. It is usual to fix this signal on the driver's left hand, similar to the running signals, irrespective of the direction in which the points lead. If the signal were placed in accordance with the direction in which the points lead, it would be placed on the driver's right hand. This is sometimes done, and there is no definite ruling as to which side of the sidings the signal should be placed, the all-important question of view often making it impossible to carry out rigidly any general regulation.

The south cross-over road should be fitted with disc (15) (17) signals in a similar manner to the north cross-over road. Where the siding and cross-over road points come very closely together the respective signals are, on some railways, placed side by side; on other railways, however, the signals are placed as near as possible to the points through which they control traffic. Theoretically it is preferable to place the signals side by side in such cases, as should there be any misunderstanding between the driver and the signalman as to the movement required, the driver might stand over one set of points waiting for the signal of the other set, whilst the signalman might pull over the first-mentioned set of points amongst the wheels of the train. If both signals are situated at the same place the driver must clear both points, and no mistake can occur.

There is a slight practical disadvantage in placing the signals together. The signal for the cross-over road being some few yards from the points makes it less easy to carry out the necessary detection of the points, as for this purpose the signal wire requires to be led from the detector at the points to the signal. In most cases the signal and detector are combined or else connected by means of a short rod.

If the cross-over road signal is placed by the side of the siding signal it would be made to detect both points. It would detect the siding points WHEN NORMAL, and the cross-over road points WHEN REVERSED in the ordinary manner, so that unless the siding points are in their correct normal position and the cross-over road correctly pulled over, it is impossible for the cross-over road signal to be cleared.

The UP LINE SIDING should be equipped with signals (19) (21) for the points (20) in the same manner as the DOWN LINE SIDING.

If these sidings are used simply for Refuging Trains, some railways fix semaphore signals for drawing out on to the main line.

WAYSIDE STATION WITH LEVEL CROSSING GATES

FIG. 224 shows a wayside station with level crossing gates. The position of the level crossing determines to a large extent the position of the signal box. If the gates are to be worked from the signal box it is essential that the box shall be built as near the crossing as possible in order to give the signalman the best possible view of the roadway. It would be possible to place the box at the other end of the platform, but in this case a special box would have to be built for controlling the gates, and in addition some system of locking would have to be introduced between the main signal box and the gate box. By placing the signal box as shown on the sketch, most of the points can be worked from the signal box; one set, however, at the west end is out of range. It will be necessary, therefore, to control this set of points by means of a ground frame interlocked with the main signal box.

Taking the "set" of the points first: There are only two pairs of points joining the UP LINE, and the siding points will be set as explained previously, but the cross-over points are complicated with a "single slip" connection to the down sidings. It is usual to take the cross-over road independently and set it as though no slips had to be considered. The slip points are then set so that they do not interfere with the use of the cross-over road. The points joining the DOWN LINE are simple cases of siding connections, and are set so that they do not interfere with the main line, and the ends of the connections joining the sidings must be set so that no vehicles can gain access to the main line. There is an alternative method of setting the slip points, as shown in Fig. 225. The ends of the cross-over road are set as previously described, but the single slip point is set so that a train backing from the UP LINE over the cross-over road would run on to the down sidings, and not on to the DOWN LINE. The end of this connection joining the down siding is of course set so that no vehicles can foul the main line. This method of setting a slip point is seldom employed, as it involves the use of an additional lever to work it; in most cases of slip points an endeavour is made to set the points so that they take the form of simple cross-over roads.

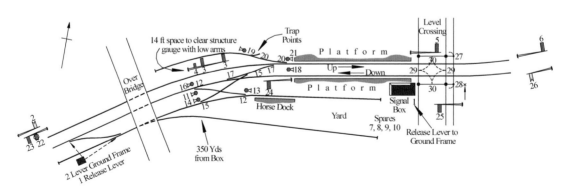

Figure 224
Wayside station with level crossing gates

Up Line signals

THE distant signal (2) will be placed at the usual distance from the home signal. The home signal must be placed clear of the cross-over road fouling point. If there is sufficient space between the siding and the main line, the signal will be placed in that

Figure 225
Alternative method of setting slip points

space, but if the space is not sufficient, the signal must be placed clear of the siding in order not to foul the structure gauge. This signal is shown with three arms on it; the top arm (3) is high to give the drivers a good view over the bridge, and the intermediate arm (3) is placed so as to be seen below the bridge (see Fig. 37). These signal arms work together.

The bottom arm (4) is a short calling-on signal to enable the signalman to indicate to the driver that he is required to draw past the signal cautiously whilst another train is standing at the starting signal.

The starting (5) signal is placed at the end of the UP platform, and assists in protecting the gates. The advanced (6) starting signal is placed a train's length ahead of the level crossing.

Down Line Signals

THE distant (26) signal is placed the usual distance from the home signal.

The home (25) signal is placed clear of the gates to protect them.

The starting (24) signal is placed at the end of the platform but clear of the cross-over road fouling point.

The advanced starting (23) signal is placed as far out as can be permitted; having regard to the signalman's view, cost of track circuit, etc., it is also placed on the wrong side of the line so that the line of sight may clear the abutments of the bridge.

A shunting (22) signal is placed lower down on the post carrying the advanced starting signal for the purpose of allowing a train to draw past the advanced starting signal when at danger, and so clearing the siding points with a long train (see p. 22).

Siding Signals

THE UP siding (20) is equipped with disc (19) (21) signals for entering and leaving, the discs being placed as previously described.

The cross-over road (17) is equipped with signals for moving in each direction, the signal (16) for moving from the DOWN to the UP LINE being placed at the fouling point of the connection to the down siding. The signal (18) from the UP to the DOWN LINE is placed opposite the signal to the UP siding, and reads both for the movement from the UP LINE to the DOWN LINE, and from the UP LINE to the DOWN Siding. This means that the signal referred to can be pulled to clear, whichever position the slip points (15) may happen to be in, and accordingly this signal does not tell the driver whether he is going to the DOWN LINE or to the DOWN Siding. In foggy weather this might lead to a collision, as, should the signalman intend to shunt a train into the siding, and should he only pull the cross-over points before clearing the signal, he might not be aware that the train was on the main line instead of in the siding.

The down siding points (12) have signals to (11), and from (13), the siding; and the siding end of the slip connection has a signal (14) from the siding to the UP LINE.

The siding connection (2) at the extreme west end of the yard is worked from the small ground frame, and as a shunter must be on the ground when a train is moved into or out of the siding it is not usual to provide signals for these movements.

Levers 29 and 30 control the crossing gate locks, and 27 and 28 the wicket locks.

DOUBLE LINE JUNCTION WITH ADDITIONAL GOODS LINE

FIG. 226 shows the signalling for a double line junction with an additional goods line. The signal box is erected as near the facing points as possible, consistent with giving the best view to the signalman and keeping the extreme points within the mechanical working.

Determining the "set" of the points

TAKING first the facing points (8) from the UP MAIN LINE to the UP GOODS LINE (south end), these must be set to resemble a common cross-over road.

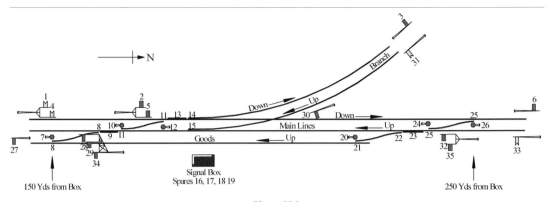

Figure 226
Double line junction with additional goods line

The cross-over road points (11) (25) are set in the usual manner. The facing points (14) from the DOWN MAIN to the BRANCH line are shown set for the MAIN line. Some railways adopt the system of making the facing points invariably lie to the left hand, the idea being that should a train over-run it would be sent to the left line, which line does not cross or foul any other line. If the train over-runs with the points set for the main line it fouls the up branch line at the diamond crossing. This latter arrangement appears to be less safe than the other arrangement, and it would be so in practice if the signalman invariably replaced all the levers after use. Under service conditions, however, when the majority of the trains run along the main line, it would entail a large number of additional lever movements if the facing points were to be moved twice for every main line train. It is very seldom that trains over-run at junctions, and when this does occur it is more likely that an over-running train would collide with a train which might be standing on the branch line than that it should meet a train coming off the branch at the crossing, as the BLOCK TELEGRAPH RULES do not allow, under ordinary conditions, of a train being accepted from the rear signal box on the branch line at the same time that another train is running along the main line with the points set for the main line.

It is the general practice to set the facing points for the most important line, as if this is not done the signalman will have the facing point lever in the "reversed" position, except when requiring to send a train to the less important line.

The trailing points (15) from the Branch line to the Up main line are also set for the most important line. It should be noted that, although the facing points *may* be set either for the branch line or the main line, the trailing points *must not* be set for the branch line *unless the facing points are also in that position.* Under no conditions must the trailing points, in a case such as this, be in a position for trains to come off the branch unless the facing points are so set that a train over-running will be sent to the branch, clear of the train coming off the branch. In practice this is ensured by the interlocking of the levers.

The points (21) (22) from the main line to the goods line (north end) are set as for a simple cross-over. The facing point end (22) on the main line could be set so as to put a train on to the goods line in case of an over-running train, but it is preferable to set this point for the main line and to allow the signalman to use his discretion as to pulling the lever in case of an emergency, or in the event of requiring to accept a train off the branch and on the up main line at the same time. The north cross-over (25) road is set in the usual manner.

Down Line Signals

AS there is a facing junction on the down line, junction distant signals are required if the line is laid out for fast traffic (see p. 9). The signal (1) for the branch line is set on a bracket to the left, the arm being lower down than the arm for the main line. The main line signal arm (4) may be on the main post, or on the right-hand side of a double bracket, according to choice. The signal post must be placed the usual distance from the home signals.

The home signals (2) (5) are a duplication of the distant signals, and must be placed clear of the cross-over road fouling point. The branch and main lines are equipped with starting signals (3)

(6), situated about a train's length ahead of the fouling point of the diamond crossing and cross-over road respectively, with due regard to the question of the signalman's view.

Up Line Signals

THE distant signal (31) for the Up Branch line must be placed the usual distance from the home signal; there will be only one arm, no signal being given for the goods line. The home signal (30) for the up branch line must be placed well back from the junction fouling point. It will be noticed that from the branch line two routes are available: (a) to the main line, (b) to the goods line. It would be possible to place two signals at this spot, one for each route. This, however, would place the dividing signals about 180 yards from the points to which they refer, and it is not desirable to have the dividing signals quite so far from the facing points, unless some device is installed to prevent the signalman from moving the points after having replaced the signal to danger, and before the train has arrived on the lock bar. In many instances it would lead to a very complicated run of signals for a driver to pick out, if the dividing signals were placed close to each set of facing points, but as a general principle dividing signals should not be placed further from the facing points to which they refer than about 120 yards, unless the above-mentioned precautions against moving the facing points are taken. The simplest system for ensuring this "holding the road" is the fixing of an intermediate lock bar (but without a facing point lock of any description attached) about half-way between the signal and the facing points, this bar being so interlocked that when the train is on this bar it is impossible to move the facing points ahead. This "holding the road" can also be effected by means of an electric lock and treadle rail contact or track circuit.

The distant signal (33) for the up main line is fixed in the same way as the corresponding signal from the branch line.

The home signal (32) for the up main line is placed at the cross-over road fouling point, and there must be a low arm (35) on the left-hand side for the goods line. Dividing signals must be placed clear of the facing point lock bar (9) of the points (8) leading to the goods line (south end). These signals cannot be fixed between the main and goods line, as the space will not permit of their erection. A main post must therefore be erected clear of the goods line, and the following signals bracketed out from it:

- the signal (28) on the extreme right must be the main line signal, and will be the highest;
- the next signal (29) must be 3 ft. lower, and reads from the main to the goods line;
- the next signal (34) placed on the main post, and spaced about 8 ft. from the last signal, is for running along the goods line. This signal is 3 ft. lower than the middle signal.

The two signals first mentioned should be spaced about 6 ft. apart and form a small group, as both read from the same point on the main line. The signal on the extreme left is spaced 8 ft. from these signals, and is thereby made to appear isolated, this being the only signal in the combination referring to trains running along the goods line.

The starting signal (27) for the Up Main line is fixed at the usual distance ahead of the last set of points.

The cross-over roads are shown equipped with signals at each end. The disc (10) signal from the Up to Down line at the south cross-over road reads either to the branch or to the main line, and would, by some Companies, be made to read also along the Up line up to disc (24) (which would give a Red light when at danger). Similarly the disc (26) signal from the Down to the Up line at the north cross-over road reads either to the main line or to the goods line, and would, by some Companies, be made to read back along the Down line up to disc (12). Should there be much shunting done over these cross-over roads, additional signals for each route might be necessary.

A disc (7) signal has been shown at the goods line end of the south connection to the same line. This signal is for backing from the goods line to the Down Main line or to the Down Branch line, picking up the application of disc (10). It would, by some Companies, be made also to apply back along the Up Goods line as far as disc (20). Similarly, the disc signal (20) shown at the goods line end of the north connection is for backing from the goods line to the Down line only.

SIGNALLING FOR STATION CONTROLLED BY TWO SIGNAL BOXES

FIG. 227 shows the signalling of a more complicated place, involving two signal boxes. This place has been signalled for *every legitimate* movement. The chief items of interest are:

(Continued on page 155)

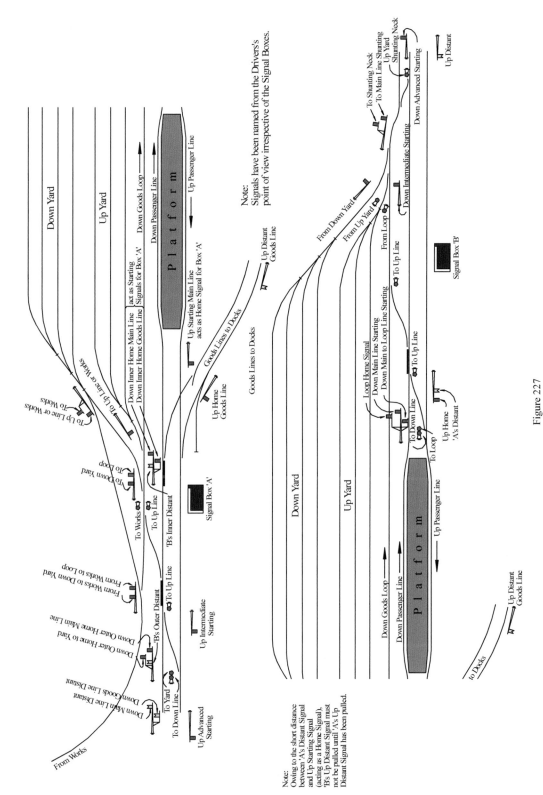

Note:
Signals have been named from the Driver's
point of view irrespective of the Signal Boxes.

Down Yard

Up Yard

Down Goods Loop →

Down Passenger Line

→ Up Passenger Line

P l a t f o r m

To Shunting Neck
To Main Line Shunting
Up Yard
Shunting Neck

Down Advanced Starting

Up Distant

From Down Yard

From Up Yard

From Loop

Down Intermediate Starting

→ To Up Line

Signal Box 'B'

Loop Home Signal
Down Main Line Starting
Down Main to Loop Line Starting

→ To Up Line

To Down Line

Up Home
'A's Distant

To Loop

Down Yard

Up Yard

P l a t f o r m

Down Goods Loop →

Down Passenger Line

→ Up Passenger Line

Up Distant
Goods Line

to Docks

Figure 227

Signalling for station controlled by two signal boxes.

Up Starting Main Line
act as Starting
Down Inner Home Main Line
Down Inner Home Goods Line
Signals for Box 'A'

Up Starting Main Line
acts as Home Signal for Box 'A'

Up Distant
Goods Line

Goods Lines to Docks

Up Home
Goods Line

Goods Lines to Docks

To Up Line or Works
To Works

To Up Line or Works

To Down Yard
To Loop

To Works

To Up Line

'B's Inner Distant

Signal Box 'A'

From Works to Down Yard
From Works to Loop

→ To Up Line

'B's Outer Distant

Up Intermediate
Starting

Down Outer Home to Yard
Down Outer Home Main Line

Down Main Line Distant
Down Goods Line Distant

To Yard
To Down Line

Up Advanced
Starting

From Works

Note:
Owing to the short distance
between 'A's Distant Signal
and Up Starting Signal
(acting as a Home Signal),
'B's Up Distant Signal must
not be pulled until 'A's Up
Distant Signal has been pulled.

154

(Continued from page 153)

- Goods lines are fitted with catch points, and the facing catch points are equipped with a facing point lock bar so as to ensure safety for *fast* goods traffic. Facing point bars would not be fixed for *slow* goods traffic, and it is not compulsory to fit any goods line points with lock bars.

- The manner in which the distant signals for both boxes are combined on posts carrying stop signals is also worth noticing.

SIGNALLING OF SIDINGS

THE signalling of sidings which are located some distance away from a station or junction, while not usually presenting any very novel or difficult problems, is a branch which, as a large proportion of private traders' sidings come into this category, warrants very careful attention. Every effort should be made to keep the cost of signalling at a minimum, consistent with safety, in order to encourage traders to apply for such sidings.

A common arrangement for sidings with a single trailing connection to a double line is shown in Fig. 228. In order to avoid the cost of a signal box, as well as the wages of a signalman, the siding is worked by a Ground Frame and no Block instruments are provided, the Block Section being from the signal box in rear to the box in advance. The Ground Frame is provided with a home and a distant signal (which stand normally "off"), but the main safeguard for a train working at the siding is the Block Telegraph. These instruments, having been placed to "Train on Line" (see Chapter 4) when the train left the rear box, will remain in this position all the time it is working at the siding and will not be altered until it has run forward past the box in advance, thus preventing (except by gross infraction of rules by two signalmen) the accepting of a following train. The home and distant signals when in their normal ("off") position ensure that the siding points are set for the main line, and, when at "danger", give some protection in case of a second train getting into the section by ignoring the signals at the rear box or by block irregularity.

In working a siding signalled in this way the train usually stops at the siding home signal, after which the engine runs ahead clear of the points with the trucks for the siding, leaving the remaining trucks and the brake van standing on the main line outside the home signal. The signals are then put back, the points pulled, and the engine propels the trucks into the siding. After picking up any trucks which are ready for despatch the engine runs out on the main line, the points are put back to normal, and the engine propels the trucks back on to the remainder of the train and couples up. The home and distant signals are then pulled "off", the Ground Frame lever padlocked, and the train resumes its journey.

With the foregoing procedure it will be seen that even while the signals are at "danger" the brake van and a portion of the train is outside the home signal, and is thus not protected by it. For this and other reasons it is now considered better practice to control this type of siding by means of an electric

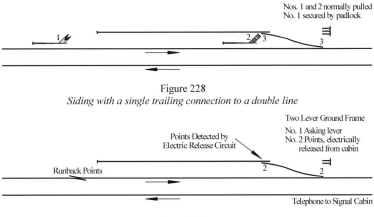

Figure 228
Siding with a single trailing connection to a double line

Figure 229
Siding with a single trailing connection with electric release

release from the rear signal box (see p. 135) as in Fig. 229. No signals are provided at the siding, but the Ground Frame control lever in the signal box is interlocked with the lever working the Starting or (if one is provided) the Advanced Starting signal. In order to ensure that the siding catch points are working effectively it is good practice to fit them with electric detection contacts which form part of the electric release circuit, so that unless the catch points are normal the electric lock in the signal box is not released.

Where conditions warrant it, additional security may be obtained if the Block Telegraph is supplemented by Track Circuiting the line from the Starting (or Advanced Starting) signal of the rear box to a short distance past the siding points, or even as far as the home signal of the box ahead, this track circuit locking the starting (or advanced starting) signal.

To prevent interference with the Ground Frame by trespassers, etc., and consequent delay to traffic it is usual to fit one of the levers with a padlock, keys for which are carried by all guards or others who operate the siding.

If with either of the foregoing siding schemes the gradients are such that a full-length train standing at the siding would have a part on a gradient steeper than 1 in 260 falling towards the rear signal box, the provision of runaway catch points a train's length back from the fouling point of the siding should be considered. In coming to a decision on this point, as in all questions of run-back catch points on gradients, the deciding factor should be not altogether the steepness at the point where the train is assumed to break away, but also the general trend of the gradients between that point and the rear signal box. Thus if the gradient is falling sharply for two or three trains' lengths and then changes to a considerable rising gradient, so that a runaway vehicle will evidently not reach the rear box, catch points are best omitted. If, however, though it appears impossible for a runaway to reach the rear box, the length of falling gradient is so great that a vehicle running back freely will attain a very high speed and be likely to leave the rails, the provision of run-back points is desirable, for the reason that if a vehicle is going to be derailed it should be derailed as early as possible while its speed is low, as in this way damage is reduced and the operation of re-railing simplified.

When a siding layout similar to the foregoing is encountered on a Single Line the ground frame is usually controlled by the single line token and runaway points—for which the necessity is decided on the same grounds as for double lines—must be operated from the ground frame and detected, preferably by electrical means. The sequence of operations in the case of a siding such as that shown in Fig. 230 would then be as follows:

When the train has come to a stand at the siding, the rear vehicle being clear of the catch points No. 3, the single line token is inserted in the lock on the ground frame (see p. 137). This releases No. 1 lever, which, when pulled, releases No. 2 lever. No. 2 lever, when pulled, withdraws the bolt

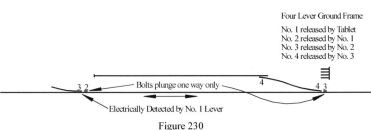

Four Lever Ground Frame

No. 1 released by Tablet
No. 2 released by No. 1
No. 3 released by No. 2
No. 4 released by No. 3

Figure 230
Siding with a trailing connection to a single line

from the trap points No. 3, and releases No. 3 lever. No. 3 lever when pulled opens the trap points and withdraws the bolt from the siding points No. 4 and releases No. 4 lever. The traffic can then be worked into and out of the siding. When this operation is complete, No. 4 lever is put back, thus enabling No. 3 to be put back. The bolt on No. 4 points plunges one way only (for the main line) and, as the rod from the ground frame to it is very short, and hence rigid, the lever cannot be forced back unless the bolt enters the notch in the stretcher, thus acting as a detector for No. 4 points. When No. 3 lever is back No. 2 lever is free to be put back. This operates the bolt on No. 3 points, but as the connection from this bolt to the ground frame is very long, no reliance can be put on it for detection purposes. No. 2 bolt and No. 3 points are therefore fitted with electrical detection contacts which control an electric lock on No. 1 lever, preventing it from being put back unless the point blades are properly closed and bolted. Until No. 1 lever is back the single line token cannot be withdrawn from the lock on the ground frame, so that the possession of the token by the engine driver is a guarantee that both the siding and trap points have been properly set and bolted.

The case of a siding on a Single Line where gradients do not call for trap points is shown on Fig. 231 at A, and is identical with the foregoing except that two levers only are required, the bolt being operated by No. 1 lever, which is released by the Single Line token. This Figure also shows, at B, a common arrangement where two connections are located close together.

In both the Double Line cases which we have considered, the siding connection was a trailing one; it sometimes happens, however, that local engineering difficulties, such as heavy earthworks or bridges, make a trailing connection

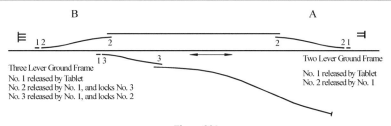

Figure 231
Sidings with connections to a single line, without trap points

very costly, and in this case a facing connection must be considered. If it is found that this will be located too far from a signal box to be worked from it by rod, the most economical arrangement is to work it from a ground frame electrically released from the box, as shown in Fig. 232. The facing points and bolt are, in this case, electrically detected by a running signal located near the points, the trap points being detected by the electric control circuit as in Fig. 229. The line is track circuited up to the clearance point of the siding on the main line, and over the trap points of the siding and a telephone is provided at the ground frame. By means of this Track Circuiting, if the train is run completely into the siding and the ground frame restored to normal, the signalman will have an indication that the main line is clear, and can allow other through traffic to pass while the engine is sorting the trucks in the siding. On lines where traffic is heavy, this facility is very useful.

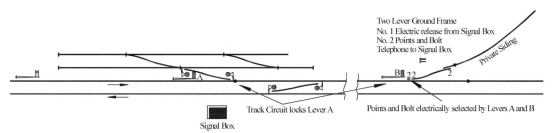

Figure 232
Siding with a single facing connection to a double line

It will be noticed that in none of the foregoing Ground Frame schemes are disc or shunt signals provided at the siding points, as it is considered that the shunter or guard operating the ground frame is so near to the points that he is able to control the movements safely by hand signals, and for the same reason the likelihood of his attempting to unbolt the points is so reduced that lock bars are not considered necessary.

SIGNALLING OF RUNNING JUNCTIONS AND WAYSIDE STATIONS

FIGS. 233, 234, 235, 236, 237, 238, give the signalling arrangements of several typical places. The running junctions and wayside stations illustrated are comparatively easy to signal. There are, however, two opposing factors which come into play in cases where high speed running is the rule, and at the same time slow traffic has to be manipulated. It is not desirable to have the signals too close together; the minimum distance should not be less than 120 yards, as, otherwise,

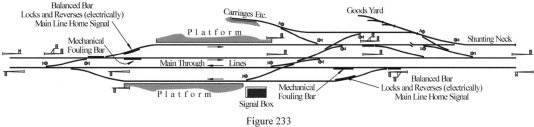

Figure 233
Signalling for double line through station.

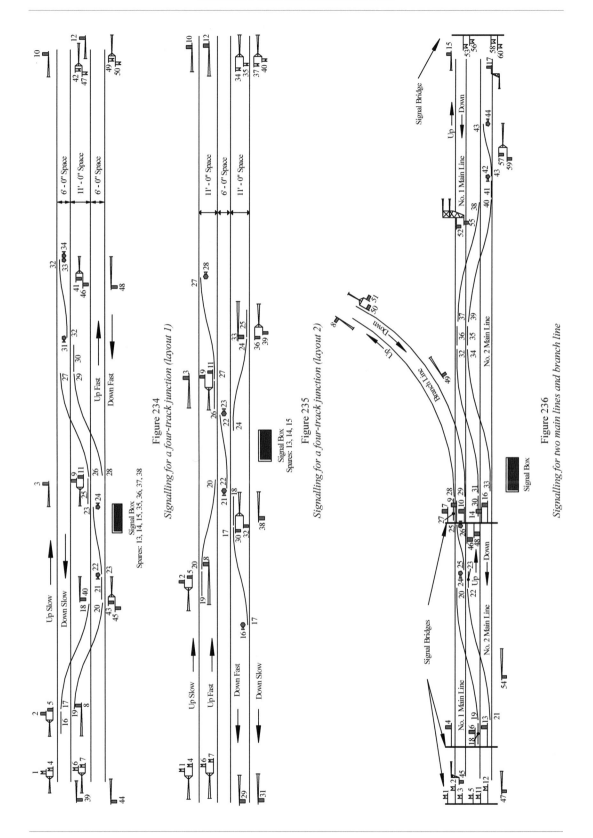

Figure 234
Signalling for a four-track junction (layout 1)

Figure 235
Signalling for a four-track junction (layout 2)

Figure 236
Signalling for two main lines and branch line

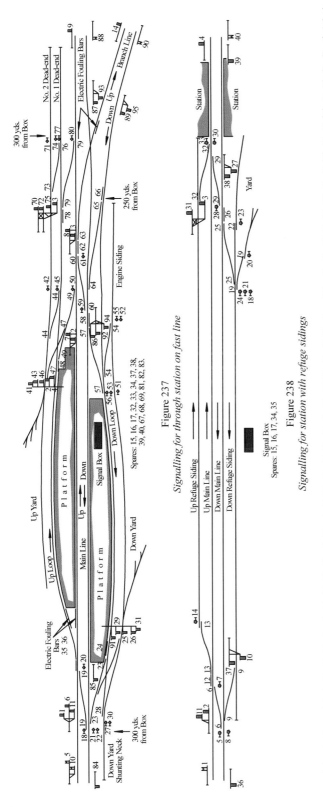

Figure 237
Signalling for through station on fast line

Spares: 15, 16, 17, 32, 33, 34, 37, 38,
39, 40, 67, 68, 69, 81, 82, 83.

Figure 238
Signalling for station with refuge sidings

Signal Box
Spares: 15, 16, 17, 34, 35.

the driver running at high speed will have difficulty in picking them out, while, on the other hand, if signals are sparingly laid out it is not convenient for shunting purposes. At places where the traffic is mixed, it is advisable fully to signal the various fouling points and place junction signals comparatively near the junction points, rather than allow the maximum distance from a protecting signal to the fouling point. In such circumstances, retaining bars, or other devices, are placed to hold the road, and the junction signals are all grouped together at one spot, but in order to accommodate high speed running, the junction signals might be duplicated at the outer home signals. Fig. 237 is signalled on this principle.

SIGNALLING A TERMINAL STATION

AT a terminal station speeds are comparatively slow, and generally there is a considerable amount of shunting from one platform line to another, for which provision must be made. If the running in and out of the platform lines only had to be considered, the most convenient arrangement would be one outer home signal at the first fouling point, and at some convenient dividing point further in to place the dividing signals, one for each platform line. Taking Fig. 239, (see page 160) the outer signal No. 2 would remain, but at the first signal bridge there would be as many signals for the main line as there are platform lines, which at this place would involve nine signals. This number of signals could not conveniently be placed all near the main line, but by using an indicator signal (see Fig. 36), this difficulty could be obviated. Placing the inner home signals at this spot would leave a great distance from the signals to the facing points near the platforms, and consequently some device would be necessary to hold the road. The simplest is to make each lock bar be released by the next bar ahead; thus No. 48 lock bar would be released by 60 or 49, 67, and No. 60 would be released by 93 or 61, 64, and so on, or alternatively the same effect can be obtained by "Bothway" locking; thus
(Continued on page 161)

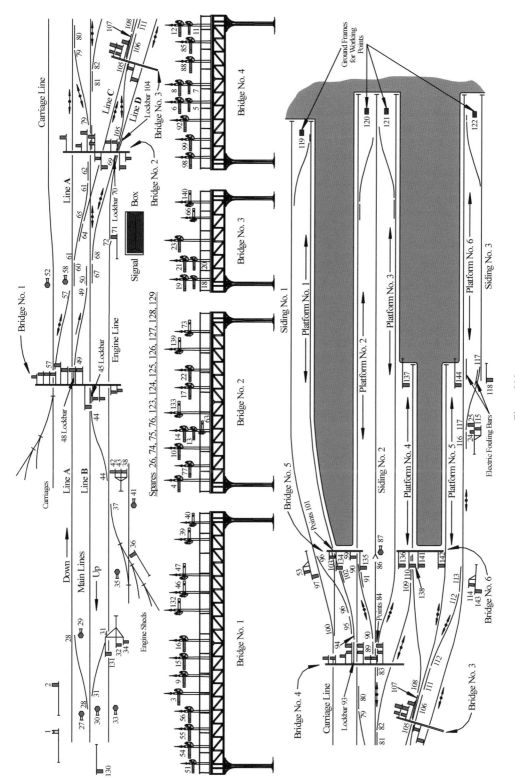

Figure 239
Signalling for a terminal station

(Continued from page 159)

48 would lock 60 Bothways, and so on. Before the signal is pulled, all the bars have to be pulled to hold the road over which the train is to be sent, and when the signal is replaced to danger it is not possible to move the lock bars near the platforms until the outer lock bars have first been put back owing to the locking, and the rear lock bars cannot be replaced until the train has passed over them. It is thus possible to unlock each facing point as the train moves clear, ready for the next required movement, but the points cannot be unlocked out of their proper order. A more expensive but safer method (as it does not rely on the lock bars holding good when forced against a running train) is the installation of Track Circuiting from the Inner Home signals up to the clearance points of the various platforms and the fitting of electric locks on the inner home signals, so arranged that when a signal has been pulled it is locked so that it cannot be put far enough back to release the bar and point back-locking (but still far enough to place the signal at danger) until the last vehicle of the train has passed off the Track Circuit. This device, however, holds all traffic back until the last vehicle of the train has safely cleared the fouling point of the platform. In a similar manner for outgoing there need only be one indicating signal placed at the end of each platform line, with an advanced starting signal placed ahead of all the point connections. In service, however, it is necessary that a terminal station shall be signalled to allow for shunting as well as running movements. This can be arranged by placing disc signals at convenient places for backing into the platform lines. It will, however, be observed that most of the lines at a terminal station are worked in both directions, so that a train running into a platform line would face these disc signals; thus, when the driver runs into the station he would be required not to observe the disc signals, but when shunting he would be required to observe them. This is an undesirable arrangement, and so it is the practice of some Companies for disc signals in situations such as this to be pulled off before a running signal is pulled.

If the station is signalled with dividing signals at reasonable intervals, little difficulty is experienced in holding the road; with slow speeds the drivers have no difficulty in picking out their signals, and no special shunting discs are required, as the signals for running can be used for all shunting purposes. It is very necessary for the expeditious working of a terminal station that all shunting movements shall be fully signalled and that there shall be no hand signalling required from the signalman. As there is no possibility of trains running away on the wrong line when backing into the platform lines, there is no need to hand-signal engines which are backing on to their trains—as is the case where there are through running lines—and calling-on signals are used for this purpose.

Fig. 239 has been signalled on the above principle, and owing to the slow speeds of trains at a terminal station, where any facing points arc more than 80 yards ahead of the dividing signal, the locking bars should be interlocked as mentioned previously.

The following is a description of all the signals at this place:

1. Distant Signal for the Down Line. This signal should be fixed at danger. If it is desired to work this signal there would be no objection to allowing it to be cleared for all the long platforms, *viz.*, Nos. 1, 2, 3, 6, but not to Nos. 4 and 5.
2. Outer Home Signal, Down Line.
3. Inner Home Signal, Down Line to Line A *(on Bridge No. 1)*.
4. Intermediate Signal, Line A *(on Bridge No. 2)*.
5. Calling-on Signal, Line A to No. 1 Platform Line *(on Bridge No. 4)*.
6. Direction Signal Line A to No. 1 Platform Line *(on Bridge No. 4)*.
7. Calling-on Signal, Line A to No. 2 Platform Line *(on Bridge No. 4)*.
8. Direction Signal, Line A to No. 2 Platform Line *(on Bridge No. 4)*.
9. Inner Home Signal, Down Line to Line B *(on Bridge No. 1)*.
10. Intermediate Signal, Line B *(on Bridge No. 2)*.
11. Calling-on Signal, Line B to No. 3 Platform Line *(on Bridge No. 4)*.
12. Direction Signal, Line B to No. 3 Platform Line *(on Bridge No. 4)*.
13. Calling-on Signal, Line B to No. 4 Platform Line *(on Bridge No. 2)*.
14. Direction Signal, Line B to No. 4 Platform Line *(on Bridge No. 2)*.
15. Inner Home Signal, Down Line to Line C *(on Bridge No. 1)*.
16. Inner Home Signal, Down Line to Line D *(on Bridge No. 1)*.

17. Intermediate Signal, Line D to Line C *(on Bridge No. 2)*.
18. Calling-on Signal, Line C to No. 4 Platform Line *(on Bridge No. 3)*.
19. Direction Signal, Line C to No. 4 Platform Line *(on Bridge No. 3)*.
20. Calling-on Signal, Line C to No. 5 Platform Line *(on Bridge No. 3)*.
21. Direction Signal, Line C to No. 5 Platform Line *(on Bridge No. 3)*.
22. Intermediate Signal, Line D *(on Bridge No. 2)*.
23. Direction Signal, Line C to Line D *(on Bridge No. 3)*.
24. Calling-on Signal, Line D to No. 6 Platform Line.
25. Direction Signal, Line D to No. 6 Platform Line.
27. Backing Signal, Up to Down Line.
29. Backing Signal, Down to Up Line.
30. Backing Signal, Up Line to Engine Line.
32. Signal from Engine Line to Up Line.
33. Signal from Dead-end.
34. Signal from Engine Line to Dead-end.
35. Intermediate Signal, to Engine Line.
36. Signal from Engine Shed.
38. Signal from Sheds, etc., to Engine Line.
39. Signal, Engine Line to Dead-end or Up Main Line *(on Bridge No. 1)*.
40. Signal, Engine Line to Engine Sheds *(on Bridge No. 1)*.
41. Signal from Engine Sidings.
42. Signal, Engine Sheds Line to Line B.
43. Signal, Engine Sheds Line to Line D *(via Up Main Line)*.
46. Signal, Up Main Line to Dead-end *(on Bridge No. 1)*.
47. Signal, Up Main Line to Engine Sheds *(on Bridge No. 1)*.
51. Signal, Carriage Sidings to Carriage Line *(on Bridge No. 1)*.
52. Signal, Carriage Line to Carriage Sidings.
53. Signal, No. 1 Siding to Carriage Line.
54. Signal, Carriage Sidings to Line A *(on Bridge No. 1)*.
55. Signal, Carriage Sidings to Line B *(on Bridge No. 1)*.
56. Signal, Carriage Sidings to Line C *(on Bridge No. 1)*.
58. Backing Signal, Line A to Carriage Sidings.
59. Signal, No. 2 Platform Line to Carriage Sidings *(via Line A)*.
63. Signal, Line B to Carriage Sidings *(on Bridge No. 2)*.
66. Signal, Line C to Carriage Sidings *(on Bridge No. 3)*.
71. Signal, Engine Line to Line D.
73. Signal, Line D to Engine Line *(on Bridge No. 2)*.
77. Signal, Line B to Line A, *via* 79 points *(on Bridge No. 2)*.
85. Signal, Line B to No. 2 Siding *(on Bridge No. 4)*.
87. Signal, No. 2 Siding to Line B.
88. Signal, Line B to No. 2 Platform Line *(on Bridge No. 4)*.
92. Signal, Line A to No. 1 Siding *(on Bridge No. 4)*.
97. Signal, No. 1 Siding to Line B.
98. Signal, Carriage Line to No. 1 Siding *(on Bridge No. 4)*.
99. Signal, Carriage Line to No. 1 Platform Line *(on Bridge No. 4)*.
103. Signal, No. 1 Platform Line to Carriage Line.
114. Signal, Line D to Carriage Sidings *(via Line C)*.
115. Signal, Line D to No. 3 Siding.
118. Signal, No. 3 Siding to Line D.
130. Advanced Starting Signal, Up Line.
131. Intermediate Starting Signal, Up Line.
132. Direction Signal, Up Line *(on Bridge No. 1)*.
133. Intermediate Signal, Line B *(on Bridge No. 2)*.
134. Starting Signal, No. 1 Platform Line.
135. Starting Signal, No. 2 Platform Line.

136. Outer Starting Signal, No. 3 Platform Line.
137. Inner Starting Signal, No. 3 Platform Line.
138. Starting Signal, No. 4 Platform Line (*via* Line B).
139. Intermediate Signal, Line D (*on Bridge No. 2*).
140. Intermediate Signal, Line C (*on Bridge No. 3*).
141. Starting Signal, No. 4 Platform Line (*via* Line C).
142. Starting Signal, No. 5 Platform Line.
143. Outer Starting Signal, No. 6 Platform Line.
144. Inner Starting Signal, No. 6 Platform Line.

Levers Nos. 119, 120, 121, 122 are control levers for the cross-over roads in the platform lines at the buffer ends (the cross-over roads being for the purpose of allowing engines to run out on the opposite platform line after their arrival).

RULES FOR SIGNALLING

THE following are the chief rules which should be observed in the signalling of any place where a considerable amount of traffic has to be operated.

Before commencing to lay out the signalling equipment ascertain the direction of running and use of every line shown on the plan and mark the direction of running with arrows.

DOUBLE LINE RULES

(1) *FIXING THE SITE OF THE SIGNAL BOX.*

First. See that it is within the M. 0. T. limit for mechanically worked points.

Second. Consider the best possible position for the signalman's view of the connections.

Third. See that there is ample space for clearing the structure gauge and for leading out the rods and wires.

(2) *DECIDING THE "SET" OF THE POINTS.*

First. See that all sidings, goods lines, and other non-block lines are trapped so that no vehicles can gain access to the passenger lines.

Second. The points inside yards and siding which do not interfere with the running lines should not, as a general rule, be connected to the signal box.

Third. Where there are slips (single or double) as far as possible set the points so that they can be coupled up like cross-over roads.

Fourth. At junctions set the facing points for the most important line, or set them so that any train over-running shall not cross the path of another train; do not, however, set the facing points so as to run a train into a dead-end siding or into a line not equipped for passenger working.

Fifth. At terminal stations see that the points are so set that a train starting without authority shall foul the fewest possible lines.

(3) *FITTING FACING POINT LOCKS.*

First. Facing point locks and bars must be fitted on all facing points on passenger lines, and should be fitted to all points over which loaded passenger vehicles are taken. (For emergency movements only, lock clamps fitted by hand are sufficient.)

Second. Facing point locks and bars should be fitted on all facing points over which high speed traffic runs even on goods lines, but especially where goods lines join passenger lines, or run close alongside them.

(4) *RUNNING SIGNALS.*

First. Place the distant signal at the correct distance from the home signals, and if this is not possible arrange for its duplication on a rear post, or prevent the rear signalman from clearing his distant signal until the signal for the box ahead has been cleared. Provide splitting distants if there are exceptional circumstances which call for them.

Second. Place protecting stop signals well clear of the fouling point to be protected. A signal should not be required to protect a fouling point more than about 500 yards ahead of it, and additional signals should be provided to prevent this distance being exceeded.

Third. At junctions provide as many splitting signals as there are routes, but as a general rule, where there are alternative routes for the same set of lines only one route is signalled.

Fourth. Junction signals should never be required to protect facing points more than about 120 yards ahead of them unless some special device for holding the road be provided, but signals for high speed running must not be placed only a short distance apart.

Fifth. Junction signals should be "stepped" to simplify their reading, the signals for the most important line being the highest, *or* the ruling adopted that the highest signal denotes the *best-running* line.

Sixth. Where signals for different lines are placed on the same bracket or bridge they should be pitched so as to form groups with wider spaces between the groups than between the individual signals of each group.

Seventh. Advanced starting signals should be placed if possible within view of the signalman; if not, the position of the signal arms as well as the presence of trains must be electrically indicated in the signal box. They should in most cases be placed about a train's length ahead of the point connections.

(5) SHUNTING SIGNALS, ETC.

First. Provide calling-on arms, where it will save the signalman waving drivers past the home signal many times per day.

Second. Provide a SHUNTING signal where it is necessary for the driver to pass the advanced starting signal for shunting purposes. This should only be the case where special difficulty prevents the Advanced Starting signal being located sufficiently far out.

Third. Provide disc signals (or miniature semaphore signals) for all legitimate shunting movements which have to be performed several times per day. Where cost is a consideration, one such signal might be used for more than one route.

Fourth. No signal must be given for running on to the wrong line unless some means can be used for putting the train eventually on to the right line, or a "Limit of Shunt" indicator is provided.

(6) *See* that no signal is placed where it will foul the structure gauge.

(7) *See* that run-away catch points are fitted on the ascending line a train's length in the rear of the home signal if the gradient is steeper than 1 in 260. (See remarks on p. 156.)

(N.B.—These rules only apply to passenger lines. For goods lines fewer signals can be provided; if, however, *fast goods* traffic has to be dealt with, it is economical to equip the running lines similarly to passenger lines.)

SINGLE LINES

FIGS. 240, 241, 242, 243 show several typical stations on single lines. Fig. 240 is a simple passing place with a goods siding; the facing points must be set so that a train in over-running will not run to the wrong platform line. The siding points are treated in the same manner as ordinary siding points on a double line.

The lock bars only lock the points when the points are normal. For instance, lock bar No. 4 is only required to be pulled when the train runs over No. 5 in the facing direction into the platform because when the train is running from the other platform with No. 13 signal pulled No. 5 is trailed over, and it is therefore unnecessary for the points to be bolted. On this account there will be only one hole in the lock stretcher, and similarly the lock stretcher on No. 11 points will only have one hole in it for the bolt to enter when the points are normal.

Figure 240
Signalling for simple passing place with goods siding

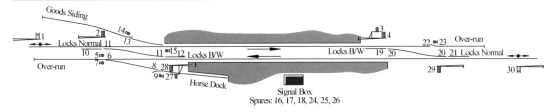

Figure 241
Signalling for passing place with goods siding and over-runs

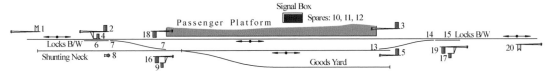

Figure 242
Signalling for single line station with goods loop

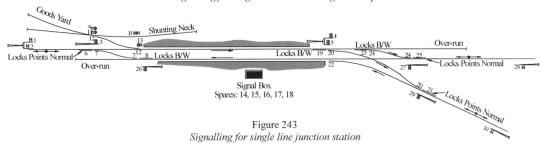

Figure 243
Signalling for single line junction station

Fig. 241 shows a passing place on a single line equipped with over-runs. A small semaphore signal is given for shunting into the over-runs, and a disc signal for backing out again. At this place the over--runs serve the double purpose of being over-runs for passing trains, and also shunting necks for working the sidings. It must be understood that they cannot serve both purposes simultaneously.

The remaining signals are similar to the previous case. The lock bars at this place are of two types; Nos. 10 and 21 lock their respective points only when the points are normal. Nos. 12 and 19 lock their respective points B/W in the usual manner. This ensures that a train in over-running No. 3 or No. 28 signal will enter the station over properly secured facing points in obedience to No. 2 or No. 29 signal. No. 2 signal must not be lowered unless No. 20 points are bolted securely. If No. 19 lock bar only locked 20 when 20 is pulled, it would involve the pulling of No. 20 points to allow a train into the platform when No. 2 is pulled. This would prevent a train from being accepted from the opposite direction, since, before a train can be accepted from the box in the rear, the points must be set for the train to run on to the correct line.

Fig. 242 shows a station on a single line with a loop line for goods traffic only. In this case dividing signals must be placed at each end of the loop line to allow trains either to the passenger platform or to the goods line. The distant signals will read for the passenger line only.

Fig. 243 is a junction station on a single line, the signalling of which needs no special mention.

Fig. 244 (page 166) shows typical cases (at A and B) of single line working of a temporary nature, such as required for bridge renewal, etc.

SINGLE LINE RULES

THE rules for single lines are very similar to those for double lines. The chief items to be noticed are:

(1) All points on a single line are FACING POINTS.

(2) On lines using the electric staff or electric tablet system, signals must be provided at staff or tablet stations, but at other stations it is *not* necessary to provide signals, the points, if any, being locked by the staff or tablet.

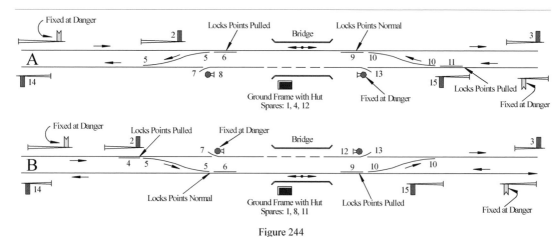

Figure 244
Signalling for single line working for engineering operations

(3) *In* setting the points at a tablet or staff station care must be taken that the train in over-running does not run on to the wrong line.

(4) *Facing* points not worked from the signal box but worked from a ground frame controlled by the staff or tablet need only be fitted with a lock bolt, no bar being required. In this case the points must be bolted when the lever is in the normal position.

NUMBERING SCHEMES

HAVING determined the signals to be provided at any place it is next necessary to determine which levers in the signal box shall operate each individual signal, set of points, or lock bar.

In an interlocking frame the levers are arranged in one line, and when facing the levers inside the signal box the left-hand lever is No. 1, and all the levers are numbered from left to right. In determining the precise number of the lever to work any particular set of points, the only ruling given by the M .O. T. is that the signalman shall have the best possible view of his connections from the position he takes up when working them. This naturally leads to all the points at the left hand end of the station, looking at them from the signalman's point of view, being worked by levers near No. 1 end of the locking frame.

Similarly the point connections at the right-hand end of the station would be worked by levers at the end of the frame having the higher numbers. It is the custom to work disc signals with levers which are situated close to the levers working the points to which they refer. For the running signals there are two general systems in use. One is to place the levers working the running signals reading from left to right at the extreme left of the frame, the levers for the running signals reading from right to left being at the extreme right-hand end. The other system is to place the running signals as near the centre of the frame as possible, the reason for this being that the block instruments are often placed in the middle of the box, and it is very convenient for the signalman to be close to his signal levers whilst giving block signals. This latter system is sometimes termed the "Grouping" System. An endeavour is made to place the levers in the interlocking frame so that the signalman shall have his levers for the various moves arranged in groups, and as far as possible pull his levers in numerical order. It is possible to do this in some small locking frames, but at junctions it becomes impracticable to provide the proper sequence of pulling for both the main and branch lines. As a rule the numbering is then made to suit the main lines, and the sequence of pulling for the branch lines then becomes irregular. An attempt is made to place the signals as near as possible to the points to which they apply, and this refers to running signals as well as shunting signals.

Whatever system is decided on, the following features are necessary for a good working arrangement:

(1) The signalman should have no unnecessary walking about the frame to pull the levers when setting the roads and clearing the signal.

(2) The order of pulling the levers should be that as far as possible the signalman only moves in one direction in setting the route.

(3) It should not be necessary for the signalman to be forced to pull or put back a lever between two levers which are already over, especially if that lever operates a heavy connection. If it is a very easy lever to work this requirement is not important.

The simplest system to remember is the one known as the Geographical System. In this system, as far as possible, the point levers work points which are situated in the same relative position outside with respect to the signal box, as the levers occupy inside the box. This ruling is adhered to irrespective of the layout of the place, and the ground disc signals are placed next to the points to which they refer, the signals reading from left to right having earlier numbers than those from right to left. The running signals are numbered from left to right, and care must be taken so to arrange them that the signalman only moves in one direction in pulling off the signals for any one route.

Fig. 226 is numbered in accordance with this system. The signals reading from left to right are taken first, and of the signals reading in this direction, those reading from the left-hand top corner to the right-hand top corner are taken first. This gives the branch line signals precedence over the main line signals at this particular place, No. 1 being the distant signal. Having numbered the branch line signals, the main line signals follow with No. 4 as the distant signal. The starting signal No. 6 finishes all the running signals from left to right. The points and relative disc signals are taken next; commencing at the extreme left hand, the disc signal from the goods line will be No. 7, the points No. 8 and the lock bar No. 9. It is very necessary that the lock bar lever should invariably come next to the lever working the points bolted by the lock bar, as every time the points require to be moved the lock bar lever is usually moved also. It should be mentioned that as a rule the lock bolt is out when the lever in the signal box is in its normal position, and to bolt the points the lever has to be pulled. Properly speaking the points should be unbolted when no train is approaching, but as this involves additional lever movements, the signalman as a matter of general practice leaves the lock levers pulled. There is no serious objection to this except at stations where a great amount of shunting is performed; in the latter case should the points be "trailed" through when they are bolted in the wrong position, much more damage would be done to the facing point gear than if the bolt were out when the mishap occurred.

The next lever will work the disc signal at the south end of the south cross-over No. 10, the points will be No. 11, and the disc from the down to the up line will be No. 12. Next come two sets of points which are opposite each other, and in cases such as this the points must be numbered to suit the sequence of pulling. Junction locking determines that the facing points at this place must be pulled before the trailing points, hence the facing points will precede the trailing points in numbering the levers. The lock bar to the facing points must come before the points themselves in order to give the correct sequence of pulling, so that the bar will be No. 13, facing points 14, trailing points 15. The north connection to the goods line would follow, but it is usual to insert some spare levers at some portion of the frame, the general practice being to insert about 1 spare to 10 working levers. In determining the exact number of spare levers required, several things have to be taken into account.

First, the possible future requirements of the place; if it is known that additions are proposed at some date in the near future, it is wise to leave sufficient spare levers for this purpose; if, however, there is no definite information as to future extensions, it is the usual custom to leave spare levers at a point where an additional siding might be put in, and in a case of this sort a group of three levers is the most convenient number to leave.

Second. It is very desirable to fix on some definite standard for the size of locking frames. A common arrangement is to commence with 5 as the unit, and only build locking frames which contain some multiple of 5, or if the design of the locking frame makes 4 a more suitable unit, that number would be chosen. This allows standard castings to be kept in stock (see p. 99). Taking 5 as the unit and counting up the number of working levers required, it is found to come to 31; this necessitates 4 additional levers to make up 35, the nearest multiple of 5. As the place now readied in numbering is as convenient as any other for leaving spare levers, Nos. 16, 17, 18, 19 are left as spares. Spare levers are not always inserted; SPACES only can be left in the locking frame ready to receive any levers should they be required. This is often done when the numbers are not reserved for a special purpose, in which case it is more economical than providing levers.

No. 20 lever will work the disc signal from the goods line (north end). It will be noticed that in the two previous cases of points which take the form of cross-over roads, one lever is made to work both ends of the connection, because in no case is one end required to be pulled without the other end. At the connection to the goods line (north end), however, two levers are used, one for each end of the points, the reason for this being that the facing point end on the main line might possibly be required to be pulled to protect a train coming from the branch line, and in that case there is no necessity for pulling the catch point end on the goods line, as this would untrap the goods line.

The lock bar will follow the facing points, and the north cross-over road with its signals come after in the usual sequence. This finishes the points, bars, and disc signals, and the running signals from right to left follow next, No. 27 being the advanced starting signal for the up main line. No. 28 is the dividing signal reading for the main line, and in order to prevent the signalman from retracing his steps in pulling the signals for a particular route, the signal from the up main line to the goods line, on the same bracket as No. 28, must follow next as No. 29. The home signal from the branch and its distant signal come next, then the home and distant signals for the main line, and finally the starting and home signals for the goods line.

The order of pulling for the various movements is as follows:

(a) To Branch Line, 14, 13, 3, 2, 1.

(b) DOWN MAIN LINE, 13, 6, 5, 4.

(c) Goods Line to Branch, 14, 13, 11, 8, 10, 7. (It will be noticed that No. 10 is out of the correct order, but this is necessary owing to the locking, which will be explained later.)

(d) Up to Down Line, 13, 11, 10.

(e) Down to Up Line, 9, 11, 12.

(f) Goods Line to Down Main Line, 25, 22, 21, 24, 20. (No. 24 is out of the correct running owing to the interlocking.)

(g) Up to Down Line (North cross-over), 25, 24.

(h) Down to Up Line (North cross-over), 23, 25, 26.

(i) From Branch Line to Main Line, 9, 14, 15, 27, 28, 30, 31.

(j) From Branch Line to Goods Line, 8, 9, 14, 15, 29, 30.

(k) UP MAIN LINE, 9, 23, 27, 28, 32, 33.

(l) Up Main Line to Goods Line, 22, 21, 23, 34, 35 (22 must precede 21 owing to the interlocking).

Fig. 236 shows the numbering arrangement for a fast-running junction with four running lines. The chief items of interest are:

(1) Where there is a single outer home signal which requires to be pulled along with two or more distant signals, all these distants must come in order before (or after as the case may be) the home signal. Nos. 1, 2, 3 distant signals being numbered before No. 4 home signal is an example of this. Where there is no single outer home signal the distants precede or follow their respective home signals. No. 5 distant preceding No. 6 home is an example of this; also 52, 53, 57, 58 show the same ruling.

(2) The same principle holds good with any signal or set of signals which have to be pulled in combination with the signals for more than one route. Nos. 8, 7, and 9, 10, are examples of home and starting signals treated in this manner.

(3) Where there are double cross-overs (usually termed "over-crossings") between running lines, allowing simultaneous movements in both up and down directions, such as the over-crossings numbered 19—20, and 21—22, the point ends should be connected to separate levers, otherwise, if ordinary junction locking is inserted, certain legitimate train movements will not be possible. This will be better understood when the interlocking of junctions has been considered.

Fig. 245 shows special cases of double and single slips and the method of numbering. In cases of this sort, while it is possible to connect several point ends on one lever, it is not advisable to connect more than three for manual working, as it makes hard pulling, and considerable trouble is experienced in maintaining all the points in good working order owing to the number of joints required in the rod connections. In all cases of connecting more than one point end to one lever great care must be taken to see that under no possible conditions can any one point end be required to be moved, independent of the others.

In all the foregoing examples, one lever has been employed to work only one signal, or one set of points, as the case may be. It is possible to work a set of points and a signal referring to those points on one lever; this is done by means of an escape crank. The first portion of the travel of the lever moves the points (there is a notch in the mid position of the floor-plate), the second portion of the lever movement does not move the points further, but it clears the disc signal. Two or more signals can also be worked by one lever,

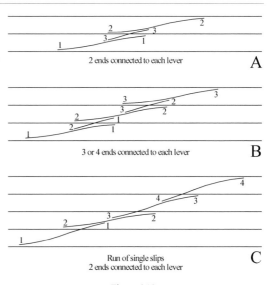

2 ends connected to each lever — A

3 or 4 ends connected to each lever — B

Run of single slips
2 ends connected to each lever — C

Figure 245
Numbering of single and double slip points

thus the disc backing *into* a siding, and the disc for coming *out* of the siding can be so worked. For this arrangement the signal lever is in the mid position normally; when it is *pushed* one signal is cleared, and when it is *pulled* the conflicting signal is cleared. As it is impossible for the lever to be in the pulled position, and right back at the same time, no other interlocking is required between the conflicting signals. At a junction where two or more signals read for diverging lines, it is possible by means of "selection" to work all the diverging signals by one lever, the position of the points determining which particular signal is cleared when the signal lever is pulled. Similarly, where two or more signals *converge* to the same line, these signals can all be worked by the same signal lever, the position of the trailing points determining which of the converging signals is cleared.

A facing point lock bar and the facing points can also be worked by one lever. Generally, however, these selecting arrangements are not favoured by signalling engineers, as it is found to be cheaper in the long run to maintain simple apparatus rather than complicated devices. Escape cranks are not desirable from a maintenance point of view, and selecting devices for manual working are also costly to keep in good working order. In power signalling, however, selecting devices are again being introduced. Apart from the difficulty of maintaining selecting devices, there is the objection that a signalman, in sending trains to diverging lines, only has to move his points, the same signal lever being used for each route. Should a train have been sent to the starting signal to stand, and a second train taken on, to go to a different destination, then, in the event of the signalman forgetting to move the facing points, the train would be sent to collide with the one already on that line. Where there is an independent lever for each signal, however, he would find that the facing points had not been set for the required route when he came to pull the signal lever for that route, the interlocking preventing him from lowering the signal until the points had been set. This could be got over by installing Track Circuiting to detect the presence of the standing train, but would be more costly to install than the independent levers for each signal.

DETECTION LIST

HAVING determined the numbering scheme, and placed the numbers on the signalling plan, it is usual to prepare a list of the detection required. This is sometimes done by putting the signal numbers near the points to be detected, the side of the points on which the numbers are placed indicating which way the points have to be set to allow the signal to be cleared. A very good arrangement is to give a small table showing which points the signals detect.

The general rule for detecting is that all the points over which the signal gives permission to pass in the facing direction should be detected. If a signal can be cleared with the points in either position, it is now usual to fit a detector, although the proper function of the detector is to check or detect the position of the switches, and if either position is right the detector can then only indicate that the point switches are close up to one side or the other, but cannot indicate in which position they are.

Taking Fig. 226 the point detection would be as follows:

Signal No. Detects.	Points when Normal No.	Points when Reversed No.
2	–	14
5	14	–
7	–	8, 11
10	–	11
12	–	11
20	–	21, 25
24	–	25
26	–	25
28	8	–
29	–	8
32	22	–
35	–	22

In some cases where the safety points of a goods line or siding is an unusual distance from the signal box, the pro-tecting signal for the main line is made to detect the safety points to ensure that they are correctly set for the run-off position. This is, however, seldom done in manual instal-lations.

It is usual to supply a reference along with a signalling plan, so that the men responsible for the execution of the work shall not be in doubt as to the purpose of any signal, or the "set" of any of the points.

In complicated schemes there is sometimes considerable difficulty in determining the best name for a signal, but nearly all running signals can be classed as distant, home, or starting. This only gives three names, and it often happens that about eight signals have to be named for one line only.

With distant signals little difficulty is experienced, and if more than one, the first signal en-countered by the driver is usually termed the Outer Distant, the second one being termed the Inner Distant. An alternative is to regard the first one as The distant signal, and term the second one the Auxiliary distant signal.

In the case of home signals, if there is no junction diverg-ing, the first one encountered is termed the Outer Home, the second one being the Inner Home. If there is a junction the Splitting signals are often termed Direction signals, as they are fixed for the purpose of indicating the di-rection in which the road is set. Sometimes the term Inter-mediate home or direction signal has to be employed where there is more than one set of such signals involved.

The term "Starting" at one time was only applied to a signal at the end of a platform, the sig-nal furthest out which gives admission into the next block section was termed the Advance or the Advanced Starting signal. It is now common to call this signal the "Starting". When leaving a ter-minal station or arriving at one, special names have to be employed to describe accurately the various signals which are required; such terms as Arrival signals, Platform Home and Platform Starting signals having to be used.

LEVER DESCRIPTION

A description to suit Fig. 226 is as follows:

No.	Lever.	Description
1	Signal	Down distant. Main to branch line.
2	Signal	Down home. Main to branch line.
3	Signal	Down advance. Branch line.
4	Signal	Down distant. Main line.
5	Signal	Down home. Main line.
6	Signal	Down advance. Main line.
7	Signal	Backing. Goods line to down line (*via* south connection).
8	Points	Facing. Set for main line. Pull for goods line.
9	Lock Bar	To No. 8 points.
10	Signal	Backing. Up to down line.
11	Points	South cross-over road. Set for main lines. Pull for shunt.
12	Signal	Backing. Down to up line.
13	Lock Bar	To No. 14 points.
14	Points	Facing. Set for main line. Pull for branch line.
15	Points	Trailing. Set for main line. Pull for branch line.
16	Spare	
17	Spare	
18	Spare	
19	Spare	
20	Signal	Backing. Goods line to down main line (*via* north connection)
21	Points	Catch. Set for run-off. Pull for goods line.
22	Points	Facing. Set for main line. Pull for goods line.
23	Lock Bar	To No. 22 points.
24	Signal	Backing. Up to down line.
25	Points	North cross-over road. Set for main lines. Pull for shunt.
26	Signal	Backing. Down to up line.
27	Signal	Up advance. Main line.
28	Signal	Up inner home. Main line.
29	Signal	Up inner home. Main line to goods line.
30	Signal	Up outer home. Branch line.
31	Signal	Up distant. Branch line.
32	Signal	Up outer home. Main line.
33	Signal	Up distant. Main line.
34	Signal	Up inner home. Goods line.
35	Signal	Up outer home. Goods line.

PLANS FOR THE MINISTRY OF TRANSPORT

SIGNALLING plans of all new works have to be submitted to the Ministry of Transport. The term "new works" includes such alterations as new junctions or sidings, which interfere with passenger lines, but not slight alterations carried out for renewal purposes, or work which is exclusively in connection with non-passenger lines. The plan submitted to the Ministry of Transport should be a print showing the lines as in Fig. 217, and give all details as to gradients, all the signals, points, lock bars, etc., with a description of the signals.

In all cases where alterations only are submitted, new work is indicated in red and old work in black, the space between the rails (4 ft. way) usually being in colour to enable the lines to be picked out easily.

11
INTERLOCKING TABLES, DIAGRAMS, ETC.

BEFORE attempting to draft up a table of locking, it is necessary to scrutinise the signalling plan (which must be numbered up) and understand the reading of each signal, also the "set" of the points, etc. It is impossible correctly to lock up a frame until the complete signalling arrangement has been mastered.

TYPES OF LOCKING TABLES

THERE are many different types of tables used for drafting interlocking. The simplest has three columns thus:

Lever No.	Released by.	Locks.
1		
2		
3		
etc.		

A very common type has four columns thus:

Lever No.	Releases.	Locks.	Released.
1			
2			
3			
etc.			

There are also different names for the columns, and the columns are sometimes transposed, the second column being "Released by", and the last column "Releases."

"Released by" is sometimes termed "Locks in reversed position", or "Backlocks". "Bothway" locks are generally inserted in the "Locks" column, occasionally in a separate column for that purpose, or else inserted both in the "Locks" column and the "Backlocks" column, with a line below the number to draw attention to its being in both columns. It is immaterial which system is employed, although with certain types of interlocking frames the four column table is preferable, but with ordinary tappet or wedge locking, where no return locking is required, the simplest table gives all the necessary information.

It must be understood that with each of the different types of tables the interlocking is drafted on the assumption that all the levers in the frame are in the normal position to commence with, and that the PULLING of the lever effects the locking. Thus the "Released by" column gives information as to the levers which require to be pulled *before* the desired lever can be pulled. It must be understood that the levers pulled to release another lever are *back*-locked when the desired lever has been pulled, so that it is not possible to return them to the normal position until the lever last pulled has been replaced.

The "Locks" column gives information as to the levers which are held in the normal position when the desired lever is pulled. "Bothway" locks are generally shown thus (4 B/W), which means that 4 is locked when normal or when pulled. It is sometimes written (B & A 4) which means before and after 4, or (4 N/R) which means 4 is locked when normal or reversed.

"Special" or "conditional" locking is usually written in the columns affected, but the condition is inserted in brackets. Abbreviations are generally employed to save space, thus:

- (4 when 7) means that 4 is locked by the lever under consideration *only* when 7 is also in the pulled position.
- (4 unless 7) means that 4 is locked by the lever under consideration unless 7 is in the pulled position, the

pulling of 7 preventing 4 from being locked.

If written in full, these abbreviations would be (4 when 7 is pulled), and (4 when 7 is not pulled), but as there is a liability of the word "*not*" being omitted in error, it is desirable to use a different word, and "*unless*" is often used for that reason. The word "*reversed*" is very generally used to denote the "pulled" position of a lever.

Another method of writing locking, used by some Companies, consists in placing a ring round a figure when the pulled or reversed position is intended, thus: ◯ and using only two columns thus :

No.	Locks.
1	3, ④ , ⑦ , 9, 11, 12, ⑫ (16 W 19) (⑱ W ⑲)

This locking would read: 1 locks 3 (normal), locks 4 and 7 reversed (that is to say—is released by 4 and 7) locks 9 and 11, locks 12 bothways, locks 16 when 19 is normal and is released by 18 when 19 is reversed.

The term *dead* lock is often applied to "direct" locking or releasing, and is used to distinguish such locks from *Bothway* or *special* conditional locking. Bothway locks are occasionally called *split* locks, or *half* locks. One lever is sometimes said to *want* another lever when it is *released* by that lever, and a lever to *precede* another lever when it *releases* that lever.

The chief rules to be observed in drafting locking are set out in Paragraph 8 of the MINISTRY OF TRANSPORT REQUIREMENTS, which reads as follows:

8. Point and signal levers to be so interlocked that the signalman shall be unable to clear a fixed signal for the movement of a train until after he has set the points in the proper position for it to pass, and bolted them as necessary, that it shall not be possible for him to clear at one and the same time any two fixed signals, which may lead to a collision between two trains, and that, after having cleared the signals to allow a train to pass, he shall not be able to move any points connected with, or leading to, the line on which the train is moving until the signal is replaced. Points also, where necessary, to be so interlocked as to avoid the risk of a collision.

In the case of signals situated at a distance from facing points, some form of locking or device may be called for to ensure that, after the signal has been passed it shall not be possible to move the facing points until the whole of the train has cleared them.

Levers operating Stop signals, which are next in advance of trailing points, operated from the same box, when worked, to lock such point levers in either position, unless this locking will unduly interfere with, and the interval between the relative signals and points is adequate for, traffic movements.

Distant signal levers must be so interlocked that the signals cannot give a Clear indication when any of the relative Stop signals are at Danger.

Electrical locking may be necessary in certain conditions between block instruments, or token instruments, and the levers operating points or signals.

Interlocking between the up and down token instruments may also be necessary on a single line where there is a block post which is not a passing place.

With track circuit in use for reminding a signalman of vehicles standing within his control, the occupation of the track should be shown by an indicator in the box, and should, when necessary, electrically lock the running signal, or signals, in rear leading on to the same line; or alternatively control the block instrument.

With automatic and controlled automatic signals and continuous track circuiting, the occupation by a vehicle of any section of track circuit should return to, and hold at, danger, a sufficient number of signals in rear to provide an adequate section of line, and may also be required to lock electrically, as necessary, in one or both positions, points on or crossing over, that section.

In addition to these Requirements there are special practices carried out by the signalling engineers of the various railway companies. Interlocking can hardly be reduced to an exact science, as there are scarcely two places precisely alike, and the traffic working, layout of the place, etc., have to be considered in deciding the interlocking. It can be taken that the main principles are the same in each case, but there can be slight variations in details.

Ministry of Transport Rules for Interlocking

THE six rules laid down by the Ministry of Transport for interlocking will now be considered one by one as they apply to a place as shown in Fig. 224 (page 150).

First Rule: *"The signalman shall be unable to clear a fixed signal for the movement of a train until after he has set the points in the proper position for it to pass, and bolted them as necessary"*.

So far as the running signals are concerned, when all the levers are in their normal position, the points *are* correctly set for the train to pass. This rule must, however, be taken to mean that any of the points being reversed must lock the running signals interfered with. Taking the points released by lever No. 1, when permission has been given for the points worked from the ground frame to be pulled, the running signals Nos. 24 and 25 must be locked. The signals Nos. 22 and 23 being ahead of the points will not be interfered with. The distant signal No. 26 must not be capable of being pulled, but this is effected by interlocking it with the running signals.

Similarly, when points No. 12 are pulled, Nos. 24 and 25 signals must be locked. When points No. 15 are pulled, both main lines are affected. In this case, however, there will be no DIRECT locking between No. 15 points and the running signals, as it will be arranged that No. 15 points cannot be reversed until the cross-over road (No. 17 points) has first been reversed; No. 17 points will be made to lock the running signals, so that as this pair of points is invariably reversed when No. 15 is required there is no necessity for No. 15 doing its own protective locking.

No. 17 when reversed must lock up Nos. 24 and 25, and in addition Nos. 3 and 4, the running signals for the other main line. Nos. 5 and 6, being ahead of the points, will not be interfered with. When No. 20 points are reversed, Nos. 3 and 4 signals will be locked. Although level crossing gates are not specifically mentioned in this particular rule, it is necessary that the gates shall lock the running signals. It is usual to effect this locking by means of the gate stop lever, or if there should be a gate lock lever, the locking would be effected by that lever. In the case under consideration, No. 30 gate stop lever will lock signals Nos. 3, 5, and 25. It will *not* lock No. 4 (the calling-on signal), as the purpose of the calling-on signal is to allow a train to draw forward cautiously to the platform whilst another is standing at the Starting Signal.

The rule also applies to the shunting disc signals. Before No. 11 can be pulled to allow a train from the main line to the sidings, points No. 12 must be reversed, that is to say, No. 11 is released by No. 12, and when No. 12 is reversed, and No. 11 pulled, it must not be possible to return No. 12 to its normal position. No. 12 is then said to be "back-locked", or locked in the "reversed position".

In a similar manner No. 13 will be released by No. 12. No. 14 will be released by No. 15. As mentioned previously No. 15 must not be reversed unless No. 17 is also reversed, otherwise the east end of No. 17 points will be damaged when the train moves in obedience to the signal. This is effected by No. 15 point being released by No. 17 points, and it is therefore unnecessary that No. 14 be released by No. 17. It should be mentioned that no positive harm would be done by inserting No. 14 released by No. 17, but unnecessary locking would be added to the interlocking frame, thus adding to the cost of the machine without gaining any corresponding advantage. No. 16 will be released by No. 17, and No. 16 must be locked by No. 15 points, to prevent their being damaged should a train move in obedience to No. 16 signal with No. 15 pulled.

No. 18 signal will be released by No. 17 points. This signal serves for two routes, therefore it must not be released by No. 15 points, neither must No. 15 points lock it. It is required, however, that when No. 18 signal is pulled, it shall not be possible to move No. 15 points, therefore No. 18 signal must lock No. 15 points bothways; that is, if No. 15 is normal when No. 18 is pulled, it will be held in that position, and if it is reversed when No. 18 is pulled, it must be held in the reversed position. No. 19 will be released by No. 20 points, and No. 21 will also be released by No. 20 points.

Second Rule: *"It shall not be possible for the signalman to clear at one and the same time any two fixed signals which may lead to a collision between two trains"*.

This is commonly termed "locking conflicting signals". Main line running signals very seldom have face-to-face conflicting signals. When No. 3 signal is pulled there is no other signal which, when pulled, authorises a train to move in a direction conflicting directly with a train moving in obedience to No. 3 signal. If No. 19 signal were pulled at the same time, two trains would be authorised to meet at the fouling point of the siding, but as No. 20 points lock No. 3 signal this is not possible. No. 19 signal if pulled at the same time as No. 21 signal would authorise a train to move out of the siding at the same time as a train moving into the siding, and this would lead to a face-to-face collision. Therefore, No. 19 must lock No. 21. Similarly, No. 11 will lock No. 13, and Nos. 14 and 16 will lock No. 18. Nos. 14 and 16, if pulled together, would allow two trains to meet at the fouling point of 15 and 17, but as No. 15 points lock No. 16 signal, and No. 14 signal is released by No. 15 points, it is obvious that 14 and 16 cannot be pulled together; therefore it is not necessary for No. 14 to lock No. 16 directly, the locking being effected indirectly.

The locking of conflicting signals is also taken to mean any two signals which, when pulled together, would give the driver conflicting orders. Thus if No. 21 signal and No. 5 signal were pulled at the same time, a driver standing at the platform would be uncertain whether he should move ahead or whether he should back into the siding. For this reason 21 must lock 5. Also 18 must lock 5, and both 22 and 23 must lock 11 and 16. For a very similar reason No. 22 must lock No. 23, as, should both signals be lowered at the same time, the driver would be instructed to move ahead only far enough to clear the siding points with the tail of his train (No. 22 shunting signal authorises this), and also at the same time to proceed into the next section (No. 23, the advanced starting signal, authorises this).

With modern arrangements No. 3 (the home signal) locks No. 4 (the calling-on signal), it being taken that conflicting instructions are given to the driver when *both* arms are lowered. On some railways No. 3 used to be released by No. 4 (see p. 19).

It can be taken as a general rule that all signals which allow trains to meet at a fouling point (converging) are indirectly locked by the point levers; at places, however, where the points are worked by hand levers fixed at the points, the locking cannot be effected in an indirect manner, and the signals must be locked directly. Thus in Fig. 246, signals Nos. 1, 2, 3, must all lock each other.

Third Rule: "*After having cleared the signals to allow a train to pass, he shall not be able to move any points connected with, or leading to, the line on which the train is moving, until the signal is replaced*".

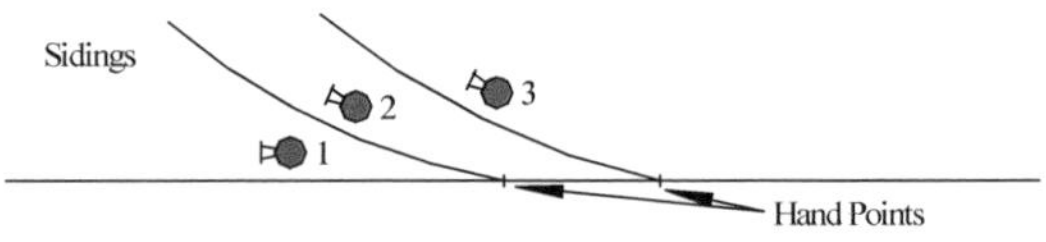

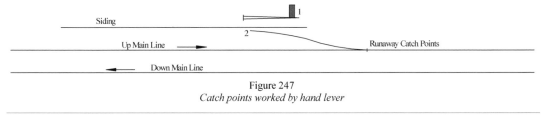
Figure 246
Conflicting signals worked by hand levers at the points

In most cases this rule simply resolves itself into the converse of Rule No. 1. There are, however, cases where the catch points of a siding or goods line are not worked by the same lever as the main line portion of the connection; thus in Fig. 247 the trailing end of the connection on the main line is operated by means of a hand lever, the points being fitted in the same manner as the run-away points on inclines. In this case lever No. 1 must lock No. 2.

Where one set of point connections is worked by two independent levers, or where an over-crossing exists, it is very important to see that this rule is not overlooked.

Fourth Rule: "*Points also, where necessary, to be so interlocked as to avoid the risk of a collision*".

The purpose of this rule is to safeguard against possible collisions when the driver moves without the authority of a "fixed" signal. In Fig. 224, if points No. 17 and No. 20 are pulled at the same time, it would allow a train to move out of the siding at the same time as a train moving from the down to the up main line. To prevent this 17 must lock 20 ; similarly, 17 will lock 12, and also 12 will lock 1.

In simple layouts a very great amount of interlocking can be saved by the judicious use of point locking, as, when points interlock, there is no necessity for the signals reading through those points to lock each other, the points affording all the necessary protection.

Figure 247
Catch points worked by hand lever

In very complicated schemes, however, great care must be taken in the point locking to ensure that legitimate simultaneous moves are not prevented.

Fifth Rule: *"Levers operating stop signals, which are next in advance of trailing points operated from the same box, when worked, to lock such point levers in either position, unless this locking will unduly interfere with, and the interval between the relative signals and points is adequate for, traffic requirements"*.

This rule means that signal No. 5 must lock Nos. 15, 17 and 20 in either position. That is, should any of these points be normal when No. 5 is pulled, they are held in that position until No. 5 has been returned to its normal danger position, and, on the other hand, should any of the points mentioned happen to be reversed when No. 5 is pulled, it is impossible to return the points to their normal position until No. 5 has been put to danger. Similarly, Nos. 22 and 23 must lock 1, 12 and 17 in either position.

This locking is commonly termed "holding the road", because whilst the signal is at clear it is not possible to move the points. The rules of the railway companies stipulate that a signal shall not be returned to the danger position until the last vehicle of the train has passed the signal.

The rule that home or starting signals in advance must lock trailing points in the rear is generally made to apply to all signals.

Sixth Rule: *"Distant signal levers must be so interlocked that the signals cannot give a Clear indication when any of the relative Stop signals are at Danger"*.

This means that No. 2 must be released by Nos. 3, 5, and 6, and No. 26 released by Nos. 23, 24, and 25.

This rule is commonly expressed as, "the distant signal to be released by the running signals ahead".

The above locking when drafted into a table appears as follows:

Lever No.	Released by.	Locks.
1	–	12, 17, 24, 25.
2	3, 5, 6	
3	–	4, 17, 20, 30.
4	–	3, 17, 20.
5	–	18, 21, 30. (B/W 15, 17, 20.)
6	–	
7	–	
8	–	
9	–	
10	–	
11	12	13, 22, 23.
12	–	1, 17, 24, 25.
13	12	11.
14	15	18.
15	17	16.
16	17	15, 18, 22, 23.
17	–	1, 3, 4, 12, 20, 24, 25, 30.
18	17	5, 14, 16. (B/W 15.)
19	20	21.
20	–	3, 4, 17, 30.
21	20	5, 19.
22	–	11, 16, 23. (B/W 1, 12, 17.)
23	–	11, 16, 22. (B/W 1, 12, 17.)
24	–	1, 12, 17.

Lever No.	Released by.	Locks.
25	–	1, 12, 17, 30.
26	23, 24, 25	
27	–	*Note.*—No locking on wickets.
28	–	
29	Gate gearing	
30	–	3, 5, 17,20, 25.
	Gate gearing Released by 30.	

The usual way in which tables are drafted is to commence at lever No. 1 and to proceed on to the last number. At a complicated place it is advantageous to decide on the *point locking first*, as this tends to prevent the insertion of duplicated locking.

EXAMPLE OF AN INTERLOCKING TABLE

CONSIDERING now a junction as shown in Fig. 226 (page 152), the locking will be as follows:

- Points No. 8 will lock points No. 22.
- Lock bar No. 9 will lock No. 8 bothways.
- Points No. 11 will lock points No. 25, unless both Nos. 14 and 22 are reversed.
- Lock bar No. 13 will lock No. 14 bothways.
- Points No. 14 will have no point locking attached.
- Points No. 15 will be released by points No. 14. (This is termed junction locking, and ensures that a train coming off the branch shall not meet a train running along the main line at the diamond crossing.) No. 15 will lock No. 25 unless No. 22 is reversed.
- Points No. 21 will be released by points No. 22, as under no conditions can No. 21 be required unless No. 22 is pulled, although No. 22 might be required in the reversed position with No. 21 in the normal position. It is not absolutely necessary that No. 21 should be released by No. 22, but by inserting this locking it saves locking No. 21 with the main line signals, thereby saving a lock.
- Points No. 22 will lock No. 8.
- Lock bar No. 23 will lock No. 22 bothways.
- Points No. 25 will lock No. 11, unless both Nos. 14 and 22 are reversed, and also lock No. 15, unless No. 22 is reversed.

The table of locking can now be commenced.

- No. 1, the distant signal, must be released by all the running signals ahead referring to its own road; that is, Nos. 2 and 3 are inserted in the "released by" column.
- No. 2 will be released by Nos. 13 and 14. In actual practice No. 14 must be pulled first, but in a table of locking it is usual to insert the numbers in their numerical order.
- No. 2 must lock the cross-over road No. 11, as if this were not done it would allow a train from another line to meet the train signalled, and in addition to this, one end of No. 11 is on the line being run over.
- No. 3, the advance signal, should lock the trailing cross-over road No. 11 in either position to obey the M.O.T. rule.

  There is some divergence of practice amongst the railways as to this. If No. 3 signal should happen to be quite close to No. 11 points, No. 3 would lock No. 11 bothways without exception.

  Where, however, No. 3 signal is a long distance from the points (it might be anything up to 500 yards), the locking is sometimes left off. If it is inserted, it is on the condition of No. 14 being reversed, but if No. 14 is normal, then the locking does not hold.

  The purpose of No. 3 locking No. 11 bothways is to prevent No. 11 from being moved until the train is clear of the points. This holding of the road can be effected by making the lock bar No. 13 lock No. 11 bothways. All trains passing over No. 11 do so in obedience to signals which are released by No. 13 lock bar, hence when the lock bar is pulled to release a signal, the points No. 11 are held in whichever position they may happen to be, and until the train has moved clear of the lock bar, it is impossible for these points to be moved. This gives all the necessary holding of the road. Of course No. 13 lock bar is fixed primarily for the purpose of holding the facing points No. 14, and that is why the M.O.T. rule does not include FACING points. On many railways, however, in order to prevent the signalman from attempting to move the lock bar whilst a train is running over it, the signals ahead lock the lock bar and facing points in both positions. With high speed traffic this locking is not irksome, but with slow moving traffic it is liable to cause delay; this will be more evident when No. 6 is considered.

If the most rigid locking is decided upon, the locking for No. 3 will be: No. 3 locks No. 12 and B/W Nos. 11 and 13 when No. 14 is reversed, and also locks No. 14 B/W.

- No. 4 will be released by Nos. 5 and 6.
- No. 5 will be released by No. 13 (this bolts the points), and will lock Nos. 11 and 14. Although No. 13 bolts the points and prevents their being moved, it is necessary to lock No. 5 with No. 14 to ensure that No. 14 facing points are in the correct position for the train to run along the main line. No. 5 must also lock No. 25. It should be noted that the junction locking (No. 15 released by No. 14) effects all the locking for trains coming off the branch, since No. 5 locks No. 14, and unless No. 14 is first pulled, No. 15 cannot be pulled to allow a train off the branch.
- No. 6 signal will lock No. 11 bothways, unless No. 14 or 25 is reversed. On either of these points being reversed, it ensures that the train signalled by No. 6 signal will *not* be trailing over No. 11 points. In other words, either No. 14 or 25 cuts out No. 11 so far as No. 6 is concerned. Similarly, No. 6 will lock No. 12 dead and No. 13 B/W, unless No. 14 or No. 25 is reversed. No. 6 will also lock No. 14 B/W, unless No. 25 is reversed, and it will invariably lock No. 26 and B/W No. 25.

Should a train be standing at No. 30 signal waiting to come off the branch, and at the same time a train be signalled along the main line with No. 6 signal at clear, it should be noticed that No. 6 locks Nos. 13 and 14 B/W, and prevents these points from being moved until the train has cleared past No. 6 signal, and also that until these points can be reversed, it is not possible for No. 15 to be set for the branch train. If the train on the main line is a very heavy mineral train, running at slow speed, it will take some considerable time before the last vehicle is past No. 6, to allow the signalman to return that signal to danger. If, on the other hand, No. 6 signal does not hold No. 14 points, then, as soon as the train is clear of the lock bar No. 13, it is possible to put back No. 13, reverse No. 14 and No. 15 consecutively, and finally pull No. 30 signal for the branch train. If this facility is allowed, it is very desirable that a fouling bar be placed clear of the diamond crossing on the main line to prevent the branch train from colliding with the main line train, should the latter not be quite past the fouling point when No. 30 is pulled. If the junction fouling point is easily seen from the signal box the fouling bar might be omitted.

- No. 7 signal allows a train from the goods line to the DOWN main line, *not* to the UP main line. It, therefore, will be released by Nos. 8, 11, and 13. This signal allows trains either to the Down main or Down branch line, hence it is not possible to lock No. 14 points; but making No. 13 release No. 7 effectively holds No. 14 points.

The road being set and properly interlocked, conflicting signals must next be locked, and as No. 12 is a directly conflicting signal, No. 7 must lock No. 12. It will also be necessary for No. 7 to lock No. 29, as No. 29 reads through No. 8 points in the opposite direction to No. 7. When a signal allows a train to proceed over a lock bar in the trailing direction, it is usual to make the signal lock the bar bothways, in order to prevent the signalman from moving the bar whilst the train is in motion, and so damaging it, and for this reason No. 7 will lock No. 9 bothways.

- No. 8 points must lock signals Nos. 28 and 34. The point locking has been previously decided, and as No. 8 locks No. 22, it is not necessary for it to lock No. 35.
- No. 9 lock bar has been made to lock No. 8 B/W (on some railways would also be made to lock No. 11 B/W).

- No. 10 signal will be released by Nos. 11 and 13; it will lock B/W Nos. 8 and 9.
  The conflicting signals are Nos. 12 (dead conflict), 28, 29, and No. 27 unless No. 8 is reversed, and these must necessarily be interlocked with No. 10.
- No. 11 points will lock signals Nos. 2, 5, 30, and 32. The point locking has already been decided.
- No. 12 signal will be released by Nos. 9 and 11, and lock B/W Nos. 13 and 14.
- Nos. 7 and 10 are conflicting signals (dead), and No. 3 is a conflicting signal when No. 14 is reversed, also No. 6 unless No. 14 or 25 is reversed, this being the converse of the locking arranged on Nos. 3 and 6.
- The locking on No. 13 has already been decided.
- No. 14 points will lock No. 5 signal.
- No. 15 points will lock No. 32 signal, in addition to its point locking.
- Nos. 16 to 19 are spare levers without interlocking.
- No. 20 will be released by Nos. 21 and 25. No. 21 being released by No. 22 makes it unnecessary for No. 20 to be directly released by No. 22.
- No. 20 will lock No. 23 B/W. No. 26 is a dead conflicting signal, and No. 34 is a conflicting signal, both of which must be locked.
- There will be no *signal* locking on No. 21.
- No. 22 will lock No. 32 signal.
- The locking on No. 23 has already been decided.
- No. 24 will be released by No. 25, and will lock B/W Nos. 8, 21, 22, 23, and 9 B/W unless No. 22 is reversed.
- No. 26 is a dead conflicting signal, and No. 28 is a conflicting signal unless No. 22 is reversed. No. 29 is also a conflicting signal under similar circumstances, but as No. 8 locks No. 22, it follows that No. 24 can lock No. 29 unconditionally. 24 should also lock 34 when 21 is reversed.
- No. 25 will lock signals Nos. 5, 32, 35, in addition to the point locking.
- No. 26 will be released by Nos. 25, 23, and if No. 22 is reversed, it must be released by No. 21 in addition. The conflicting signals are Nos. 20, 24 (dead), and No. 6.
- No. 27 will lock B/W No. 8, and also B/W No. 9, unless No. 8 is reversed (if the bar is not regarded as sufficient for this purpose).
  The only conflicting signal is No. 10 when No. 8 is normal.
- No. 28 will be released by No. 9 and will lock No. 8. It will lock No. 10 dead, B/W Nos. 11 and 15 unconditionally, B/W No. 22, unless No. 11 or 15 is reversed, and will lock No. 24, and B/W Nos. 23 and 25, unless No. 22 is reversed.
- No. 29 will be released by Nos. 8, 9, will lock B/W Nos. 11, 15, 23, 25, and will lock Nos. 7, 10, 24 dead. The conditional locking is not needed on Nos. 24, 25, because No. 8 locking No. 22 would make it inoperative.
- No. 30 will be released by Nos. 9, 15, and will lock No. 11. There is no conflicting signal.
- No. 31 will be released by Nos. 27, 28, 30, the distant signal not being pulled for the goods line.
- No. 32 will be released by Nos. 9, 23, and lock Nos. 11, 15, 22, 25.
- No. 32 has no conflicting signal.
- No. 33 will be released by Nos. 27, 28, 32, the distant signal not being cleared for the goods line.
- No. 34 needs no releasing, but locks No. 8 dead, B/W No. 21, the only conflicting signal is No. 20. For very strict locking No. 34 should lock No. 24 and B/W Nos. 23, 25, when No. 21 is reversed.
- No. 35 will be released by No. 21 (No. 21 is released by No. 22) and No. 23. It will lock No. 25; as No. 22 locks No. 8, it will not be necessary for No. 35 to lock No. 8 directly.

The locking table will appear as follows:

Lever No.	Released by.	Locks.
1	2, 3	
2	13, 14	11.
3	–	[12 (B/W 11, 13) when 14] (B/W 14).
4	5, 6	
5	13	11, 14, 25.
6	–	26 (B/W 25) (B/W 14 unless 25) [12 (B/W 11,13) unless 14 or 25).
7	8, 11, 13	12, 29 (B/W 9).
8	–	22, 28, 34.
9	–	(B/W 8).
10	11, 13	12, 28, 29 (27 unless 8) (B/W 8, 9).
11	–	2, 5, 30, 32 (25 unless 14 and 22).
12	9, 11	7, 10 (3 when 14) (6 unless 14 or 25) (B/W 13, 14).
13	–	(B/W 14).
14	–	5.
15	14	32 (25 unless 22).
16	–	
17	–	
18	–	
19	–	
20	21, 25	26, 34 (B/W 23).
21	22	
22	–	8, 32.
23	–	(B/W 22).
24	25	26, 29 (28 unless 22) (34 when 21) (B/W 8, 21, 22, 23) (B/W 9 unless 22).
25	–	5, 32, 35 (11 unless 14 and 22) (15 unless 22).
26	23, 25 (21 when 22)	6, 20, 24.
27	–	[10 (B/W 9) unless 8] (B/W 8).
28	9	8, 10 (B/W 11, 15) [24 (B/W 23, 25) unless 22] (B/W 22 unless 11 or 15).
29	8, 9	7, 10, 24 (B/W 11, 15, 23, 25).
30	9, 15	11.
31	27, 28, 30	
32	9, 23	11, 15, 22, 25.
33	27, 28, 32	
34	–	8, 20 (B/W 21) [24 (B/W 23, 25) when 21].
35	21, 23	25.

If the facing point lock bar is used to hold the points in the rear of it, the locking will be as follows:

Lever No.	Released by.	Locks.
1	2, 3	
2	13, 14	11.
3	–	12 when 14.
4	5, 6	
5	13	11, 14, 25.
6	–	26 (B/W 25) (12 unless 14 or 25).

Lever No.	Released by.	Locks.
7	8, 11, 13	12, 29 (B/W 9).
8	–	22, 28, 34.
9	–	(B/W 8, 11).
10	11, 13	12, 28, 29 (27 unless 8) (B/W 8, 9).
11	–	2, 5, 30, 32 (25 unless 14 and 22).
12	9, 11	7, 10 (3 when 14) (6 unless 14 or 25) (B/W 13, 14).
13	–	(B/W 11, 14).
14	–	5.
15	14	32 (25 unless 22).
16	–	
17	–	
18	–	
19	–	
20	21, 25	26, 34 (B/W 23).
21	22	
22	–	8, 32.
23	–	(B/W 22, 25).
24	25	26, 29 (28 unless 22) (34 when 21) (B/W 8, 21, 22, 23) (B/W 9 unless 22).
25	–	5, 32, 35 (11 unless 14 and 22) (15 unless 22).
26	23, 25 (21 when 22)	6, 20, 24.
27	–	(10 unless 8).
28	9	8, 10 (24 unless 22) (B/W 15).
29	8, 9	7, 10, 24 (B/W 15).
30	9, 15	11.
31	27, 28, 30	
32	9, 23	11, 15, 22, 25.
33	27, 28, 32	
34	–	8, 20 (24 when 21) (B/W 21).
35	21, 23	25.

With some types of interlocking frames it is not desirable to insert very much "special" or "conditional" locking. In most of the above cases the conditions could be omitted, but at a busy place it would hamper traffic. Thus 3 could lock 12 dead, but if this were done, then, when a train is standing at 3 waiting for the block section to be cleared, while at the same time a train is backing from the main line through the cross-over road 11 with 12 pulled, it would be impossible to pull off 3 to allow the train to depart until signal 12 has been put back to danger. If a long train was being shunted, it might mean a considerable delay to the train at No. 3 signal. In order to avoid delay the signalman might put No. 12 signal to danger, but this would be infringing the rule prohibiting the replacement of a signal until the train is clear of the points protected by that signal *(Rly. Cos. Rule, No. 61)*.

The insertion of the condition that 3 only locks 12 when 14 is reversed removes this difficulty.

Having written out the locking table, it is usual to check it through by cross-checking all the locks, thus 2 locks 11, the converse must hold good, *i.e.,* 11 must lock 2; again, 3 locks 12 when 14, therefore 12 must lock 3 when 14, and so on right through the locks. There is no corresponding column for checking the "Released by" interlocking, but as a rule this locking is simple, and is checked by referring to the signalling scheme to verify the accuracy of the numbers. Similarly there is no column for the converse of the B/W locks; these also have to be verified by reference to the numbered scheme.

Duplicated locking can often be detected by carefully going over the table and following out the "Released by" column, and should any of the levers releasing a particular lever have inserted against them the same locking as the lever released, such locks can be removed from the latter lever. Sometimes, however, locking is purposely duplicated, such as inserting locking on the points, and also on the signal released by the points, but only in cases where it is anticipated that the locking on the points might prove to be a hindrance to the traffic; so that should the point-locking be dispensed with, the locking is already on the signals without further alterations or additions.

MISCELLANEOUS CASES OF INTERLOCKING

IN the locking for Fig. 224 (page 150) it will be noticed that both 24 and 25, the running signals, lock the points ahead. Should there happen to be a third running signal in the rear of the points it also, as a general rule, would lock the points. It is the general ruling that ALL RUNNING SIGNALS lock the points ahead. Under exceptional circumstances, however, to facilitate traffic working, it is regarded as sufficient if two running signals in the rear are locked with the points; especially is this the case where slow speeds are enforced, such as entering a terminal station.

All points leading to the same line are, as a general rule, interlocked ; thus, in Fig. 224, 17 locks 12 and 1. In Fig. 226, 22 locks 8, but under exceptional circumstances, in order to facilitate traffic working, the point interlocking can be relieved, and in this case inserted on the signal No. 35, as there is a running signal (No. 34) between the points concerned. Thus in Fig. 224, if there had been a signal at the fouling point of No. 1, referring to the main line, the locking between 17 and 1 might be taken out, but in that case 18 signal must lock 1.

Where it does not *seriously* interfere with the traffic, it is always advisable to adopt the more rigid ruling for locking points and signals.

The foregoing ruling as to locking back for all the running signals applies very well at high-speed junctions, and at such places should also be made to apply to holding facing points as well as locking trailing points. This often means a considerable amount of special locking at compli-cated stations to ensure that the road is held right through when the outer signal is pulled. Taking Fig. 236 (page 158), the outer signal No. 4 will be released by No. 27 lock bar to hold facing points No. 28, and, in addition, unless 28 points are reversed, No. 4 signal must be released by No. 36 lock bar to hold No. 37 facing points. Similarly, No. 13 will be released by No. 18 lock bar to hold No. 19 points, No. 30 lock bar to hold No. 31 points, and if 31 is reversed it must also be released by No. 36 lock bar; 13 also must be released by 32 points when 31 is reversed, in order to complete the setting of the route.

At stations where high speeds are not allowed it is sufficient to hold the road only about 200 yards ahead of the next signal at danger, and if there should be a second signal within this dis-tance the locking need not be taken further ahead than that signal. At a terminal station, where speeds are necessarily very slow, this ruling can safely be adopted.

In Fig. 236, junction point locking is inserted between 19 and 21, between 20 and 22, and so on with the other similar cases; 37 will be released by 38, and 40 will be released by 39, but such interlocking cannot be inserted between 19 and 20, as although when No. 6 signal is pulled both 19 and 20 are required, yet 19 must not be released by 20, because 20 is, in return, released by 22, and if 22 is released by 21 it would prevent a train running along the main line with 54 and 6 both in the clear position.

Points No. 37 and 40 being released by 38 and 39 respectively do not interfere with legitimate movements, as the trailing points, which in this case *precede* the facing points, are not themselves released by any other points.

Similarly, 31 must not be released by 32, because 32 in turn must be released by 34; 29 will be released by 28, but 31 must not be released by 29, because it would prevent signals 9 and 55 from being pulled at the same time; this, of course, would prevent a movement which is quite legiti-mate, and accordingly would be a serious hindrance to the traffic. In all these cases, where the protecting locking cannot be inserted on the points, it must be inserted on the signals. Thus, as 19 must not be released by 20, it is necessary to make 46 lock 19, and, in addition, the locking must be carried further back, so that both Nos. 49 and 52 must lock 19 unless 22 is reversed.

The reversing of 19 point and the clearing of No. 4 signal simultaneously is objectionable, but if No. 4 signal locks No. 19 points it prevents the legitimate movement of 4 and 48 being pulled at the same time, since 48 will be released by 21, 22, 23, and as 21 is released by 19, therefore 19 must not lock 4.

Under no conditions must point interlocking be allowed to interfere with legitimate movements of trains, and for that reason some engineers prefer to dispense with the ordinary junction locking at places such as this, and leave it to the signalman to set his facing points as he may think best under the varying conditions of his traffic. Thus if a train is accepted along the No. 1 down main line with 22 and 21 points pulled, it would be possible for No. 19 points either to be normal or reversed as best suits the signalman; thus if a second train should be approaching on the No. 2 up main line No. 19 points could be left normal, so that an over-running train would not travel across and foul the No. 1 main lines. This freedom, however, is hardly necessary, since the fouling point of the No. 2 up main line with the 21-22 over-crossing is much nearer than the fouling point of the 19-20 over-crossing and the No. 1 up main line. In addition to this, the Block Telegraph Rules prevent the acceptance of trains which will foul each other at junctions and over-crossings. It is preferable in all cases to insert the junction locking between junction points, as it considerably relieves the signal and point interlocking. For instance, signal 52 locks 38 points—the points giving direct access to the line which 52 controls—and 52 also locks 39 points—the points over which a train runs in obedience to that signal. The fact that junction locking is inserted saves all consideration of signals which would allow trains to foul the line protected by 52, because 39 being locked no trains can foul the line *via* 37 points, and similarly no trains can foul the line *via* 40 points.

Referring to Fig. 238 (page 159), the locking required by the following levers is of interest. Signal No. 5 reads through the cross-over road 6, and may run a train along the main line or the refuge siding. If No. 13 points are reversed and 32 points also reversed, 33 signal must be locked. The complete locking on 5 is:

Lever No.	Released by.	Locks.
5	6, 12	7, 14, 36 (33 when 13) (B/W 32 when 13).
The locking for No. 6 will be:		
6	–	2, 9, 11, 37, 38 (29 unless 13 and 25) (32 unless 13) (39 unless 25).

The locking for No. 8 signal is also interesting. It will be released by 9 to let trains into the down refuge siding. As the train proceeds along the refuge siding it first encounters No. 19 points. These must be locked B/W and the conflicting signal 20 locked. The next points, viz., 25, must be locked B/W, but if 25 points are reversed it would allow the train (assuming it over-runs No. 24 signal) on to the wrong line; therefore, to prevent this, when 25 is reversed No. 8 must be released by 29 points to put the train on to the right line, and then must lock the conflicting signal No. 30. The complete locking for No. 8 signal is:

Lever No.	Released by.	Locks.
8	9 (29 when 25)	10, 20, 23, 36 (30 when 25) (B/W 19, 22, 25) (B/W 26 when 25).

No. 33 signal must be treated in a similar manner, its locking being:

Lever No.	Released by.	Locks.
33	32 (6 when 13)	4, 31 (5 when 13) (B/W 13) (B/W 12 when 13).

The outer signal No. 39 will lock 6 and 9 unless 25 is pulled, the complete locking for No. 39 being:

Lever No.	Released by.	Locks.
39	26	29 (6, 9 unless 25) (B/W 9 when 25).

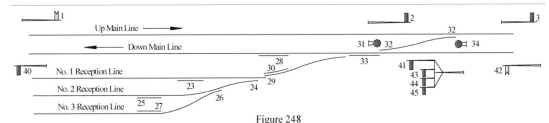

Figure 248
Reception lines with fouling bars worked from signal box

Fig. 248 shows some reception lines with fouling bars at the fouling points of the various lines, these bars being worked by levers in the signal box. It is necessary so to interlock the signals, or points, and the fouling bars, as to compel the signalman to move the fouling bars and thus to ensure that no vehicles are standing in the way before a train is let into one of the reception sidings.

If the signals leading into one particular siding are made to be released by the fouling bars, it is certainly necessary to pull the bars concerned before the signal can be pulled, but, having once pulled the bar, there is nothing to prevent the signalman leaving the bars in the pulled-over position and so not utilise the bars at all. To obviate this it is necessary to compel the signalman to put the bars back when the signal has been restored to danger.

The simplest method to treat a case of this sort is to make the signal for entering the siding lock the bar for that siding. This then ensures that the bar referring to the siding being used is normal, and before a train can be run into an adjoining siding, in obedience to the proper signals, the bar must be pulled to release the required signal, thus making certain that the first train has moved up past the fouling point. It is, of course, no guarantee that the train will not creep back and foul the other lines *after* the bar has been moved.

If it is required to allow for a possible backing movement of the train, then an electric fouling bar must be provided.

The complete locking for a place arranged as in Fig. 248 is as follows:

Lever No.	Released by.	Locks.
1	2, 3	
2	–	32.
3	–	34 (32 B/W).
23	–	43 (34 when 29 unless 24).
24	23	
25	–	44 (34 when 29, 24 unless 26).
26	24, 25	
27	–	45 (34 when 26, 29).
28	–	41 (34 unless 29).
29	28	
30	29	
31	32	34 (40 unless 29) (B/W 28, 29, 33) (23, 24, 25, 26, 27, 30 when 29).
32	–	2, 41, 43, 44, 45.
33	–	(B/W 29, 32).
34	32, 33 (25 when 29 unless 24) (27 when 24, 29 unless 26) (30 when 29)	3, 31, (23 when 29 unless 24) (25 when 29, 24 unless 26) (27 when 26, 29) (28 unless 29) (B/W 24, 26 when 29).
40	–	(31 unless 29).
41	33	32, 28.
42	40, 41	
43	25, 30, 33	23, 32.
44	24, 27, 30, 33	25, 32.
45	26, 30, 33	27, 32.

The complete sequences of pulling for the signals are as follows:

Lever No.	Sequence of Pulling
1	3, 2.
2	–
3	–
32	32.
34	32, 33; or 25, 28, 29, 30, 32, 33; or 23, 24, 27, 28, 29, 30, 32, 33; or 23, 24, 25, 26, 28, 29, 30, 32, 33.
40	–
41	33.
42	40, 41.
43	25, 28, 29, 30, 33.
44	23, 24 27, 28, 29, 30, 33.
45	23, 24, 25, 26, 28, 29, 30, 33.

It will be noticed the numbering of the levers is not strictly in accordance with the "Geographical System"; they have been slightly rearranged in order to give the best possible sequence of pulling.

Fouling bars can also be interlocked with the signals by using rotation locking as described on p. 106. Thus it would serve the required purpose if, say, 44 were released by 23 and 27, with the addition of the requirement that 44 *having once been pulled and put back, cannot be pulled a second time until 23 and 27 have been put back and pulled again.*

This form of locking is not very desirable, and is not inserted if the required effect can be obtained without it, but locking of this description is occasionally employed in connection with home and starting signals, or inner and outer home signals, as previously described on p. 107.

SINGLE LINE EXAMPLES

THE locking for single lines varies but slightly from that of double lines.

Consider Fig. 240 (page 164). Apart from the lock bars locking the facing points one way only, the chief variation is that No. 2 must lock 14.

The block telegraph rules for single lines prohibit both home signals being cleared at one and the same time, and to prevent this rule from being broken, the interlocking referred to is inserted. The starting signals in these cases must *not* hold the facing points in the rear B/W, otherwise it may hinder 3 and 13 being pulled at the same time, which is required to send both trains away simultaneously. It must also be noted that with single-line places like Fig. 240 it is *not* possible to set the road throughout before lowering the signals; thus No. 2 signal must *not* be released by 11 points.

The complete table for Fig. 240 is:

Lever No.	Released by.	Locks.
1	2, 3	
2	4	7, 14.
3	11	8 (B/W 7).
4	–	5.
5	–	4.
6	7	8.
7	–	2. 14. *
8	7	3, 6.
9	–	
10	–	

Lever No.	Released by.	Locks.
11	–	12.
12	–	11.
13	5	
14	12	2, 7. *
15	13, 14	

* Locks marked thus may be omitted, but 6 should then lock 14.

Fig. 241 (page 165) is a passing place with over-runs, and in this case it is possible to set the road ahead of the home signals more in accordance with the methods adopted on double lines. When over-runs are provided it is not usual to make the home signals lock each other, there being no danger should one train over-run. If, however, the over-runs are short this locking would be inserted. The block telegraph rules, however, only allow the concession of slowly moving past the home signal into the station instead of bringing the train to a dead stop, and require that only one train at a time shall be allowed into the station. For this reason some companies maintain the locking between the home signals even where there are adequate over-runs. Leaving out this locking, Fig. 241 will be locked as follows:

Lever No.	Released by.	Locks.
1	2, 3.	
2	10, 19	14, 23.
3	19, 20	15 (B/W 14).
4	19, 22	15, 23 (B/W 14).
5	6	20, 27, 29 (B/W 12, 21).
6	–	8, 11.
7	8	9.
8	–	6 (29 unless 11).
9	8	7.
10	–	11.
11	–	6, 10 (23 unless 14).
12	–	(B/W 11).
13	14, 19	15, 23.
14	–	**2.**
15	14	3, 4, 13 (B/W 19, 20, 22).
16	–	
17	–	
18	–	
19	–	(B/W 20).
20	–	5, 21, 22.
21	–	20.
22	–	20.
23	22	2, 4, 13 (11 unless 14) (B/W 10, 14, 19).
24	–	
25	–	
26	–	
27	6, 12	5.
28	11, 12	
29	12, 21	5 (8 unless 11).
30	28, 28	

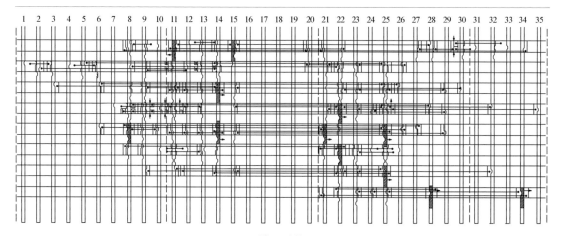

Figure 249
Diagram of locking

LOCKING DIAGRAMS

FOR shop purposes a diagram of locking is prepared, giving all the necessary information for the erection and interlocking of the required frame. For a new pattern apparatus standard drawings must, of course, be made, but after these have been made, diagrams only (not to scale) are supplied for future locking frames. Symbols of varied character are employed for the various standard parts, each railway and manufacturing firm having their own distinctive signs.

For tappet locking frames it is usual to draw the locking plungers with the required notches, locks and connecting rods, showing with dots which locks are to be fastened to any particular connecting rod. With channels suited for five connecting rods, the first line shows the outside TOP connecting rod, the second line shows the corresponding BOTTOM connecting rod, the third line shows the centre TOP connecting rod, the fourth line shows the second BOTTOM one, and the fifth line shows the last TOP connecting rod. This notation is quite easy to remember, as if the centre lines of the connecting rods were shown in plan they would take up the positions as described.

With "table" locking, as described in Fig. 168, the diagram is drawn out with No. 1 lever at the left-hand end of the diagram. Fig. 249 shows a locking diagram for a locking frame to suit a junction as shown in Fig. 226 (see locking table on p. 181).

The sketch of the lines as numbered up is sometimes given for the information of the locking fitters. It is very convenient for all concerned to have the sketch of the lines, etc., on hand when testing the interlocking; at the same time it gives the men in the works an opportunity of knowing the reason why the locking is inserted between certain levers, and enables them to carry out their portion of the work in a more intelligent manner.

Fig. 250 shows a diagram of locking to suit a locking frame, as illustrated in Fig. 165, with sketch and table complete ready for the shops. The diagram is drawn up on the assumption that the fitter *always* faces the levers in the same manner as the signalman, whereas in practice should any locking be placed on the reverse side of the levers next to the front wall of the signal box, it is necessary for the locking fitter to work with the diagram upside down in order to make the diagram agree with the actual locking. To prevent this the locking on the reverse side of the levers is sometimes numbered from the fitters' stand-point.

A very convenient method of laying out the locking in diagram form is to have a standard gauge sheet made up with lines ruled on it giving the position of each locking plunger, and also of each channel for the connecting rods and locks. The sheet is used in conjunction with a sheet of tracing paper by having the gauge sheet and tracing paper fixed on the drawing board. The notches in the locking plungers, locks, connecting rods, etc., can then be drawn in with pencil, the gauge sheet keeping the work evenly spaced.

When all the locking has been laid out in its final position it can then be traced on to tracing cloth, and blue prints taken for the shop.

In laying out the interlocking from a locking table, it is preferable to commence with the levers having the most locking on them, also "special" or "conditional" locking, and any locking which

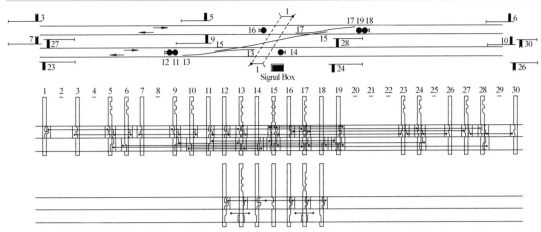

No.	Released by.	Locks.	No.	Released by.	Locks.	No.	Released by.	Locks.
1		5, 9, 13, 17, 24, 28	11	15	19, 23	21	Spare	
2	Spare		12	13	14, 15, 23	22	Spare	
3	5, 6		13		1, 9, 24	23		11, 12 (13, 15 B/W)
4	Spare		14	13	10, 12, 15	24		1, 13
5		1, 17	15	13, 17	12, 14, 16, 18	25	Spare	
6		18, 19 (15, 17 B/W)	16	17	15, 18, 27	26	23, 24	
7	9, 10		17		1, 5, 28	27		16 (17 B/W)
8	Spare		18	17	6, 15,16	28		1, 17
9		1, 13	19	15	6, 11	29	Spare	
10		14 (13 B/W)	20	Spare		30	27, 28	

Figure 250
Diagram of locking with sketch and locking table

may be between levers a great distance apart; the smaller pieces can be inserted conveniently after the more important runs have been laid down. For a beginner it is useful to lay out the locking roughly on a sheet of ruled foolscap paper, using each line to represent a lever. Having laid all the locking out separately in this manner, it can be examined to see which runs of locks can most conveniently be grouped together in the same channel. In many cases it will be found that two or more levers lock another group of levers—these can be coupled up to the same connecting rod; thus if 5, 7, 20 lock 3, 17 (B/W 21, 22, 23), all this combination would be grouped on one connecting rod. Where one group of levers locks another group, and in addition each member of the first group also individually locks certain other levers which are not locked by any other lever in the same group, it is sometimes possible to insert the individual and separate locking on independent connecting rods, and then, by making all the independent connecting rods butt against another rod (the latter being connected to the locks of the second group), the whole locking can be effected in one channel.

Fig. 251, A, (overleaf) shows such an arrangement.

With tappet locking a great saving of material can be effected by the judicious use of butts. If several levers each lock one another by the insertion of butts and cutting the connecting rod, the locking can be effected in the space occupied by one connecting rod.

As a rule connecting rods should not be made to butt against each other *unless* the butt can be placed below a locking plunger; if this cannot be done, it is preferable to rivet a block on each connecting rod and allow the blocks to butt.

A very common use for butts is where the signal from a siding is released by the points, and locks the ingoing signal; at the same time the ingoing signal is also released by the same points, and locks the outgoing signal. Fig. 251, B, (overleaf) shows the common way of arranging this.

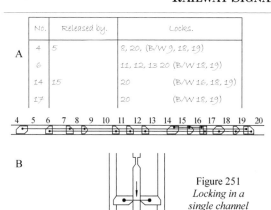

No.	Released by.	Locks.
4	5	8, 20, (B/W 9, 18, 19)
6		11, 12, 13 20 (B/W 18, 19)
14	15	20 (B/W 16, 18, 19)
17		20 (B/W 18, 19)

A

B

Figure 251
*Locking in a
single channel*

Generally all long connecting rods should be arranged so as to work in tension; if long rods are in compression there is a danger of their buckling unless they are well guided. Even then the locking between the levers feels "springy", which is very undesirable, as a springy lock is liable to suggest that the locking is sticking, and that it could be pulled if sufficient force were applied. With long rods in tension this never occurs.

It is generally found convenient to design the locking with the numbered signalling scheme on hand with the table, as this allows the draughtsman to see at a glance which levers are likely to run in groups for interlocking purposes. For locking frames of ordinary size it is convenient to lay out what is purely point locking by itself and then treat each signal lever separately, combining the connecting rods by means of butts when two or more signals lock the same group of points, etc. The designing of interlocking gives much scope for ingenuity in compressing the locking into the smallest number of channels. It often happens, however, that it is not ultimately economical to compress the interlocking too much, as where the locks and rods are crowded together very tightly it adds to the cost of manufacture and makes the apparatus more difficult to maintain. It is preferable from every point of view to spread out the locking slightly as, although this may use a trifle more material, it involves less time in making, and in the event of alterations being required to the locking frame when in service, it costs less to add a slight amount of locking where room can be found for it in the channels already in use.

TESTING INTERLOCKING

AFTER the locking frame has been erected in the signal box the interlocking must be tested to ensure that no errors have been made either in designing the locking or in its execution.

The operation of testing is best carried out by working entirely from the Diagram of Signals hung up in the box.

Suppose the locking frame to suit Fig. 226 (page 152) is to be tested. First see that all the levers are in the normal position, then try to pull lever No. 1; this should be locked. To release it pull 2, then try 1, which should still be locked; then pull 3, and try 1, which should now be free to be pulled. Having pulled 1, try to restore 2 and 3 to their normal position; they should be back-locked. The test so far has ensured that 1 cannot be pulled unless 3 has first been pulled, also that 1 back-locks 2 and 3; it has not, however, ensured that 2 must be pulled before 1 can be pulled, as should 1 lock 2 B/W by error, the mistake would not have been revealed by the above test. To ensure that 2 must be pulled to release 1, put 1 back, also 2 back, *leaving* 3 *over*, then try 1; if it cannot be pulled it is certain that 2 being normal is the cause. In any releasing combination it is necessary to have all the members of the group pulled to test the back-locks, then put back each one by itself, having the others over to test each individual lock.

To test the locking on No. 2: First try it with all the levers normal; it should be locked; then pull 14 and 13, when it should be free; having pulled it, try to put back 13 and 14, which should be back-locked. Then try 11, which should be locked. Put back 2, pull 11 and then try 2. This ensures that 11 is actually being locked by 2 and not by 13 or 14, or that it requires releasing before it is free to be pulled. Next put 11 back, also 13 and 14, then pull 13 and try 2, which should be locked; this proves that 14 must be pulled before 2. Put back 13 and pull 14, and try 2, which should be locked. This proves that 13 must be pulled before 2.

To test the interlocking on 3. This lever is free to be pulled without any other levers first being pulled. Pull 3, try 11, 13 and 14; the latter only should be locked. Put back 3, pull 14, then pull 3 again, try 11 and 13, which should now be locked (unless the locking is arranged as per the table on p. 182), try back 14, which should be back-locked. Put back 3 and pull 11 and 13, then pull 3 again, when 11 and 13 should be back-locked. Put back 3 once more and pull 12—this signal requires 9, which must first be pulled—then try 3, which should be locked; put back 12 and pull 3, when 12 should be locked.

The testing of the other levers is precisely similar to the above. Where there is an outer home signal such as No. 4 in Fig. 236 which can be pulled for 3 routes, each route must be tested separately both for releasing and locking.

GENERAL RULES FOR TESTING INTERLOCKING

THE general rules for testing are:

1. FOR A DISTANT SIGNAL.—Pull the necessary home and starting signals, next try the back-locks, and then test each individual release in the group.

2. FOR OTHER SIGNALS.—Set each possible route complete, pull the signal and then try to break the route by endeavouring to put back each of the levers which have been pulled. Next put back the signal lever and set each item of the route in the wrong position one by one, and see that the signal lever cannot then be pulled. Next with the correct route set, pull over, one at a time, any point levers which would enable any vehicle to foul the line on which a train would run in obedience to the signal, and see that the lever is locked, and conversely pull the signal lever and see that the above-mentioned point levers cannot then be pulled. Next, with the lever normal pull any conflicting signal and test the lock, and then the converse lock should be tested. If there are any trailing points in the rear, test the bothway locking.

3. FOR POINTS.—Any releasing should be tested as described for a signal. Interlocking between points must be tested by pulling the point lever and seeing that all points leading on to the same line are locked; also the converse locks must be tested in the same manner as for a signal.

4. FOR LOCK BARS.—See that the facing point lever concerned cannot be moved when the lock bar lever is pulled, unless the bar when normal locks the points, in which case the converse must be tried.

5. FOR ANY CASE.—When testing the locking between any two levers see that each lever is free to be pulled when the other one is in its correct position.

LEVER COLOURS AND RELEASING NUMBERS

TO assist the signalman in picking out his levers, the levers are painted different colours, according to the uses they have to serve (Fig. 252 overleaf). The usual lever colours are:

> Distant Signals .... YELLOW
> Other Signals .... RED
> Points .... BLACK
> Lock Bars and Bolts .... BLUE
> Spare Levers .... WHITE

One or two companies paint *all* their signals Red, whether distant or stop signals.

Release levers (such as for ground frames) are painted BLUE top half, BROWN lower half.

Levers with locking attached, which are temporarily spare levers, but which require to be worked by the signalman, are by some companies painted the lower half WHITE, the top portion being painted whatever colour they are ultimately intended to be.

> Fog Machine Levers .... BLACK AND WHITE CHEVRONS
> Gong Levers .... GREEN
> Gate Stop Levers .... BROWN
> Wicket Gate Levers .... BROWN

There is no definite ruling applicable to all railways for these minor levers, each company having its own standard colours.

To assist the signalman in the sequence of pulling his levers, RELEASING NUMBERS are painted on the sides of the levers. (Some companies have special plates bolted to the levers for this purpose.)

The numbers are painted on the lever in the exact order in which each lever has to be pulled before the required lever can be pulled, and in this respect the releasing numbers differ from the "RELEASED BY" column of the interlocking table. With distant signals it is generally considered sufficient to give only those signal levers which immediately precede it, but with other signals,

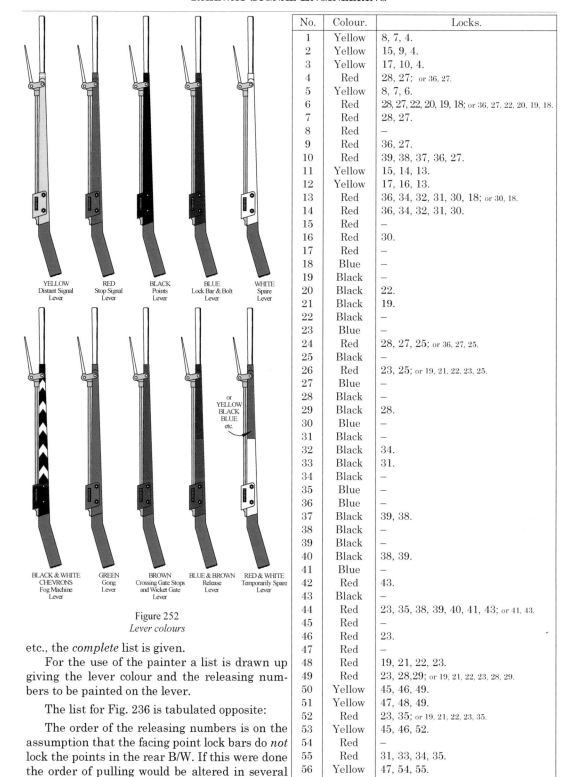

YELLOW
Distant Signal
Lever

RED
Stop Signal
Lever

BLACK
Points
Lever

BLUE
Lock Bar & Bolt
Lever

WHITE
Spare
Lever

BLACK & WHITE
CHEVRONS
Fog Machine
Lever

GREEN
Gong
Lever

BROWN
Crossing Gate Stops
and Wicket Gate
Lever

BLUE & BROWN
Release
Lever

RED & WHITE
Temporarily Spare
Lever

or
YELLOW
BLACK
BLUE
etc.

Figure 252
Lever colours

No.	Colour.	Locks.
1	Yellow	8, 7, 4.
2	Yellow	15, 9, 4.
3	Yellow	17, 10, 4.
4	Red	28, 27; or 36, 27.
5	Yellow	8, 7, 6.
6	Red	28, 27, 22, 20, 19, 18; or 36, 27, 22, 20, 19, 18.
7	Red	28, 27.
8	Red	–
9	Red	36, 27.
10	Red	39, 38, 37, 36, 27.
11	Yellow	15, 14, 13.
12	Yellow	17, 16, 13.
13	Red	36, 34, 32, 31, 30, 18; or 30, 18.
14	Red	36, 34, 32, 31, 30.
15	Red	–
16	Red	30.
17	Red	–
18	Blue	–
19	Black	–
20	Black	22.
21	Black	19.
22	Black	–
23	Blue	–
24	Red	28, 27, 25; or 36, 27, 25.
25	Black	–
26	Red	23, 25; or 19, 21, 22, 23, 25.
27	Blue	–
28	Black	–
29	Black	28.
30	Blue	–
31	Black	–
32	Black	34.
33	Black	31.
34	Black	–
35	Blue	–
36	Blue	–
37	Black	39, 38.
38	Black	–
39	Black	–
40	Black	38, 39.
41	Blue	–
42	Red	43.
43	Black	–
44	Red	23, 35, 38, 39, 40, 41, 43; or 41, 43.
45	Red	–
46	Red	23.
47	Red	–
48	Red	19, 21, 22, 23.
49	Red	23, 28, 29; or 19, 21, 22, 23, 28, 29.
50	Yellow	45, 46, 49.
51	Yellow	47, 48, 49.
52	Red	23, 35; or 19, 21, 22, 23, 35.
53	Yellow	45, 46, 52.
54	Red	–
55	Red	31, 33, 34, 35.
56	Yellow	47, 54, 55.
57	Red	23, 35, 38, 39, 40, 41.
58	Yellow	45, 46, 57.
59	Red	41.
60	Yellow	47, 54, 59.

etc., the *complete* list is given.

For the use of the painter a list is drawn up giving the lever colour and the releasing numbers to be painted on the lever.

The list for Fig. 236 is tabulated opposite:

The order of the releasing numbers is on the assumption that the facing point lock bars do *not* lock the points in the rear B/W. If this were done the order of pulling would be altered in several instances, thus:—26 would be 25, 23, or 19, 21, 22, 25, 23; the bar 23 having to be pulled after all the points have been pulled.

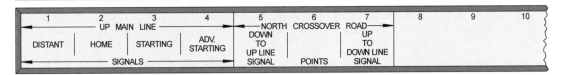

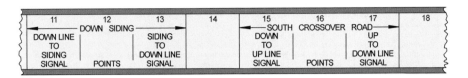

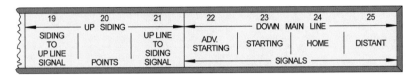

Figure 253
Lever direction board

DIRECTION OR NAME BOARD

EITHER a name plate or a direction board is supplied on which the levers are described. If name plates, they are fastened to the levers. Usually, however, a board is placed at the back of the levers, giving the description of each lever. The lettering is, in most cases, painted on, and the colour sometimes corresponds with the colour of the lever to which the lettering refers; but it is common for the lettering either to be black letters on a white ground or the reverse.

Fig. 253 shows a sketch of a name board to suit the place illustrated in Fig. 223.

The board is about 12 in. wide, and the lettering for each lever is placed immediately behind that lever.

DIAGRAM OF SIGNALS AND POINTS

IN addition to the direction board, a diagram of all the lines and signals is made and hung in the signal box. Fig. 254 (p. 194) shows a reduced copy of a diagram for a junction station, the numbering being on the grouping system. In practice the space corresponding to the 4 ft. way is coloured according to the use to which the line is put. Diagrams are usually made to some standard size, and the following dimensions, etc., are about the average practice.

Sizes for diagrams (overall size of paper):

2 ft. 3 in. by 15 in.

3 ft. by 18 in.

4 ft. by 21 in.

Sizes of details on diagrams:

4 ft. spaces, $5/16$ in.

6 ft. spaces, $7/16$ in.

11 ft. spaces, $9/16$ in.

Point circles, $3/4$ in. diam.

Signal Posts, from 1 $3/4$ in. upwards, according to height of post on ground.

Signal Arms, $3/8$ in. by $3/4$ in.

The colours of the lines differ considerably on different railways, the most common being:

MAIN PASSENGER LINES—Blue, Yellow, or Neutral Tint.

SLOW PASSENGER LINES—Same as above, but lighter tint.

Goods and Mineral Lines (Running)—Sepia or Purple.

Goods and Mineral Lines (Reception)—Neutral Tint or Green.

(*Continued on page 195*)

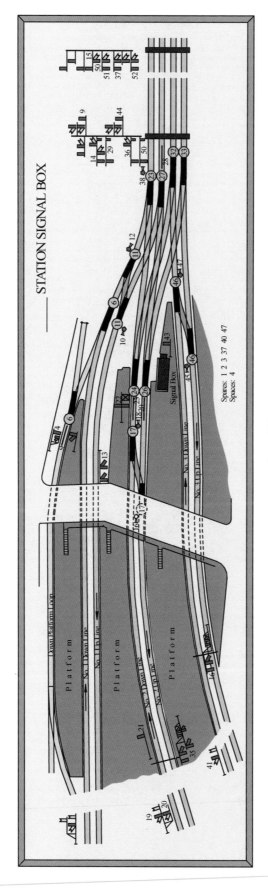

STATION SIGNAL BOX

Figure 254
Signal box wall diagram of signals and points

(Continued from page 193)

Sidings and Cross-over Roads—Blue or Sienna.

The point circles are often tinted either Sienna or Yellow.

Station Platforms—Sienna or Yellow.

Roadways—Sienna or Yellow.

On many railways diagrams are now made on the "Single Line" pattern, that is, with one heavy line for both rails, similar to the signalling sketches, the "set" of the points being indicated in the same manner. Where diagrams are made as in Fig. 254, the "lie" of the points is indicated by shading them, or blocking them in.

SIGNAL NOTICES

IN order to inform the drivers, etc., as to the reading of the signals, and to advise them of any alterations in the signalling arrangements, a signal notice is issued. The signal notice must be issued either by the operating superintendent of the line or by the general manager. Usually it is issued by the former officer. The signal notice may be issued weekly, and have all the engineering alterations for the week included therein, or a separate notice may be issued dealing only with signals at irregular intervals when required. Where new or altered signals require describing they are described as being on a certain side of the line, so many yards in a certain direction from the signal box, and the reading of the signal is given. If a new signal box is to be brought into operation it is described with respect to the side of the line on which it is built, and its distance from the nearest signal box on either side of it, and from the nearest mile post.

If two or more signals are on one post, such as home and distant signals, or junction signals, it is usual to give a diagram of the signal, numbering the arms, and then giving a description of each arm. At very complicated stations or junctions, where extensive alterations have been carried out, a complete signalling diagram, giving all the lines, etc., is included in the signal notice.

A signal notice to suit Fig. 223, assuming that a new signal box has been built to replace an old box, and that an additional cross-over road has been provided, is appended overleaf.

Where extensive alterations are in progress, it is very often necessary to erect the signals some time before the work is ready for bringing into operation, and in such cases it is usual to fix a cross on the arms of signals which are not in use, as shown in Fig. 255, A. Ground disc signals not in use are either covered with a bag, or the target (or miniature semaphore as the case may be) is removed.

Where there is a large number of junction signals to be brought into operation it is very convenient for the drivers in learning the signals if the reading of the signals is painted on the arms (see Fig. 255, B).

This is sometimes done by pasting a printed slip on the arm, but it should not be of a permanent character, and, in any case, it is of no assistance for night working.

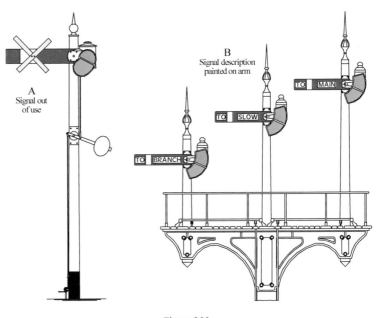

Figure 255
Crossed and lettered signal arms

GENERAL MANAGER'S DEPARTMENT

S I G N A L   N O T I C E

_____ day, the _____ day of _____ 19____.

_____ STATION SIGNAL BOX.

From _____ A.M. until _____ P.M. (or until the work is finished) the Engineering Department will be engaged altering the connections at the above place.

◊ THE EXISTING signal box will be dispensed with, and the working of the points and signals will be transferred to a new signal box, situated on the down side of the line at 103 miles 1,594 yards, and about 75 yards northward from the existing signal box.

◊ ALL SIGNALS will stand at *Danger,* and flagmen will be provided by the Engineering Department to conduct traffic under the signalman's instructions.

◊ A new cross-over road between the up and down main lines about 230 yards north of the signal box will be brought into use. The new signalling arrangements will be as follows:

1. UP DISTANT SIGNAL.—A single armed signal fixed by the side of the up line about 1,150 yards north of the signal box.
2. UP HOME SIGNAL.—A single armed signal fixed by the side of the up line about 235 yards north of the signal box.
3. UP STARTING SIGNAL.—A single armed signal fixed between the up siding and the up main line about 10 yards north of the signal box.
4. UP ADVANCED STARTING SIGNAL.—A single armed signal fixed by the side of the up line about 670 yards south of the signal box.
5. DOWN DISTANT SIGNAL.—A single armed signal fixed by the side of the down line, about 900 yards south of the signal box.
6. DOWN HOME SIGNAL.—A single armed signal fixed between the down main line and the down sidings about 120 yards south of the signal box.
7. DOWN STARTING SIGNAL.—A single armed signal fixed at the end of the down platform about 166 yards north of the signal box.
8. DOWN ADVANCED STARTING SIGNAL.—A single armed signal fixed by the side of the down line about 400 yards north of the signal box.
9. SHUNTING SIGNALS.—A Ground Disc signal fixed at the points on the down main line about 230 yards north of the signal box, which will read:—Backing, DOWN MAIN to UP MAIN.
10. A GROUND DISC signal fixed at the points on the up main line about 168 yards north of the signal box, which will read:—Backing, UP MAIN to DOWN MAIN.
11. A GROUND DISC signal fixed at the points on the down main line about 10 yards south of the signal box, which will read:—Backing, DOWN MAIN to DOWN SIDING.
12. A GROUND DISC signal fixed at the catch points of the down siding about 52 yards south of the signal box, which will read:—From DOWN SIDING to DOWN MAIN.
13. A GROUND DISC signal fixed at the points on the down main line about 51 yards south of the signal box, which will read:—Backing, DOWN MAIN to UP MAIN.
14. A GROUND DISC signal fixed at the points on the up main line about 115 yards south of the signal box, which will read:—Backing, UP MAIN to DOWN MAIN.
15. A GROUND DISC signal fixed at the catch points of the up siding about 115 yards south of the signal box, which will read:—From UP SIDING to UP MAIN.
16. A GROUND DISC signal fixed at the points on the up main line about 170 yards south of the signal box, which will read:—Backing, UP MAIN to UP SIDING.
17. RUN-AWAY CATCH POINTS.—Catch points have been fitted on the DOWN MAIN LINE about 500 yards south of the down home signal.

12
DOUBLE WIRE WORKING

THE operation of points and signals by double wires has only been mentioned incidentally in the preceding chapters because this method has been quite foreign to British practice; and rodding was, prior to 1925, specified in the Ministry of Transport Requirements for the mechanical operation of points.

As, however, this method has been in extensive use in certain Continental countries for many years, and appears likely to become popular in some of the British colonies, and may possibly find a useful application in suitable locations in this country, it is felt that the present volume would not be complete without a description of its principles. In this connection it may be noted that "other approved method" has been mentioned as an alternative to rodding for the mechanical operation of points in the latest editions of the Ministry of Transport Requirements.

GENERAL

THE method of operating signals by ordinary single wire, it will be recalled, is for the wire to be attached to a lever in the cabin. When the lever is pulled, a pull is transmitted to the wire, which in turn, working around wheels, through pulleys and so out to the signal, exerts a pull on the balance lever of the latter, and by overcoming the weight of the spectacle and counterweight (if any is provided), moves the arm to the clear position. When pulled off the signal is held in that position by the catch on the locking frame lever, which engages with the notch in the floor plate, all the operating wire being in tension.

Similarly, point blades when operated in the usual way by rodding are held against the stock rail by the force exerted by all the rodding between the points and the box, this force again depending on the engagement of the locking frame lever catch with the notch in the floor plate.

With double wire operation, on the contrary, the signals and points are operated and held in position by cam wheels located at the function concerned, so that the operating wires are only actively employed when the point or signal is in the act of being moved.

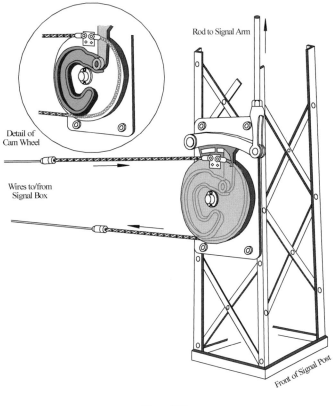

Figure 256
Double wire signal cam wheel

SIGNALS

THE operation of a signal cam wheel can be followed from Fig. 256 (see previous page) , in which it will be seen that when the wheel is rotated in a clockwise direction, the roller attached to one extremity of the T-lever which operates the signal rod will first traverse a portion of the cam groove lying close to the centre of the wheel and circumferential to it, the T-lever thus not being moved. As the wheel continues to rotate, the roller will enter the portion of the cam groove which inclines out towards the rim of the wheel, and thus the T-lever will be raised, so clearing the signal. A further rotation of the wheel causes the roller to move through a portion of the groove lying along the circumference of the wheel, which again does not move the T-lever, while a final rotation would cause the roller to follow the groove back to the centre of the wheel, thus restoring the signal to danger. This last portion of the cam groove is not required for normal operation, and its use will be described later. The portions of the groove which are circumferential and thus do not move the T-lever are intended to give a certain amount of latitude or "escapement" to the operation of the wheel.

POINTS

THE arrangement of a point operating cam wheel (see Fig. 257) is based on the same escapement principle, except that the roller is in this case attached to the wheel while the cam is formed on the point operating lever.

Facing points can be operated by an identical type of cam wheel (see Fig. 258). The plunger in this case is connected to the wheel itself in such a way that it is rapidly withdrawn by the initial "free" rotation of the wheel.

When the latter has rotated to

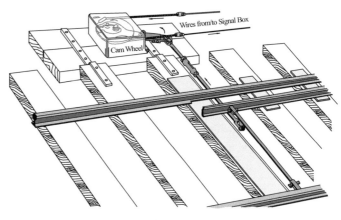

Figure 257
Double wire trailing point cam wheel layout

the position where its roller takes up the drive to the points, the drive to the plunger is almost on dead centre and so only moves slowly for a small distance while the points are being moved. When the wheel escapes from the point drive, the plunger drive again comes into an effective position and the plunger is rapidly forced home.

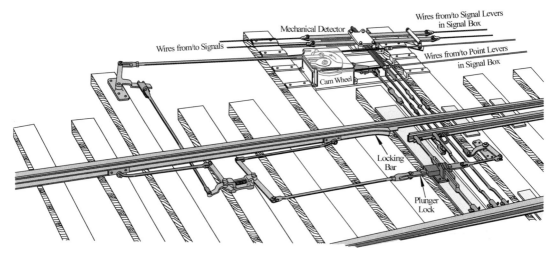

Figure 258
Double wire facing point cam wheel layout

COMPENSATION

TWO wires are attached to each cam wheel, and this is caused to rotate when the tension in one wire is greater than that in the other, and so, as the wheel "escapes" or is released from its drive to the points or signal except when the function is in the act of being moved, it will be seen that at all other times the two wires to each wheel have the same tension.

It is this feature of "equal tension in each wire except when the lever is being operated" which renders the compensation of double wires practicable. Wire is subject to the same expansion and contraction from temperature changes as rodding, and cannot, like rodding, be worked in compression in order to compensate for these changes.

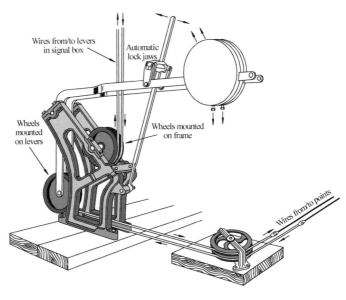

Figure 259
Double wire compensator

A different principle is therefore employed for compensation. As the two wires to any cam wheel are of equal length and run close together they expand and contract equally, and so the temperature changes cannot tend to move the wheel. Some means, however, must be found for taking up slack when the wires expand and letting it out again when they contract, and for this purpose the Compensator (of which a type is shown in Fig. 259) is used. This consists of two L-shaped levers mounted side by side in a frame, each carrying a wheel at one end and a heavy weight at the other. The wires to the cam wheels are led round wheels mounted on the frame and round the wheels on the L-levers in such a way that the weighted levers tend to put the wires in tension and so take up the slack. So far the apparatus merely consists of a pair of what are commonly termed "jockey pulleys", such as are frequently found on chain or belt drives. If, however, the L-levers were left free to move, it will easily be seen that an attempt to move a cam wheel by tightening one wire and slackening the other would merely result in one weighted L-lever moving upward and the other downward, thus taking up uselessly the wire travel needed to operate the cam wheel. To prevent this, various types of devices are employed, all with approximately the same effect, *i.e.*, that of coupling the two L-levers so that they are free to move together, but become locked if forces are applied which tend to make them move apart. The device shown in Fig. 259 consists of a pair of jaws, one carried on each L-lever, between which passes a rod secured to the frame of the compensator. So long as the two L-levers move together the jaws slide freely up and down the rod, but if one L-lever attempts to move in the opposite direction to the other the jaws are drawn together, so gripping the rod and locking the levers. Thus it is that with temperature variations (in which both L-levers tend to rise or fall together) the L-levers are allowed to move freely, while operating forces (which tend to move the L-levers in opposite directions) cause the jaws to lock, so that the full travel of the wires is transmitted to the cam wheel.

LOCKING FRAME

THE foregoing describes all the apparatus which is essential to double wire operation, but as the simultaneous movement of the two wires can with great convenience be effected by levers of the "Turnover" type, this class of frame has come to be almost invariably associated with double wire operation.

One lever of a "Turnover" type frame is illustrated in Fig. 260. The lever is comparatively short, about 2 ft. 6 in., and its fulcrum is about 3 ft. above the signal box floor. The angular movement is 180°, so that a very long lever travel is obtained, the top of the lever moving a distance of about 7 ft. 9 in. This big angular movement makes possible an exceedingly simple and effective type of catch handle locking mechanism. The locking tappets are operated by cranks pivoted on the frame, the opposite ends of the cranks being operated by extensions from the catch rods. The pins coupling the cranks to the catch rods are located so that when the catch handles are raised the pins are brought into line with the lever fulcrum. Thus the effect of lifting the catch handle of a lever which is normal (as in Fig. 260) is to cause the tappet to move downwards about 1 in. to the "mid-stroke" position. While the lever is being moved from normal to reverse, the locking crank pin is in line with the lever fulcrum, and so the crank and tappet are not moved. When the lever is fully reversed and the catch handle released, the locking crank pin is moved upwards, causing the tappet again to move down about 1 in. to the full "reverse" position.

Speaking mathematically, the work done in operating a lever is measured by the force required to move the lever multiplied by the distance through which it is moved. So that for the same amount of work less force is required to be exerted on a lever when its travel is increased.

In applying this principle to the design of locking frames, it has to be borne in mind that the strength required to operate a lever is governed

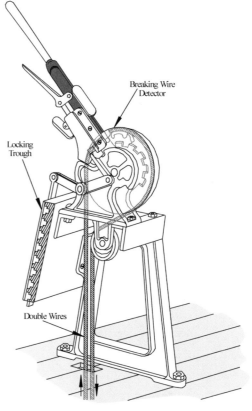

Figure 260
Double wire turnover locking frame

not only by the force which is required, but also by the suitability of the lever's position for the application of the force. Thus with ordinary types of locking frames, though the force required would be reduced by an increase in the lever travel, a travel greater than about 4 ft. 6 in. would bring the lever when fully normal or reverse into positions which would be very awkward for operation, and would more than counterbalance the benefit gained from the reduction in the force required. In addition to this, the virtual impossibility of reaching a lever between others which were in the pulled position would render very difficult the numbering of a frame with this excessive travel.

In the case of a Turnover frame, however, the levers when in the reversed position are out of the operator's way, so that in numbering up the frame, pulling between levers need not be considered.

With ordinary locking frames having a lever travel up to 4 ft. 6 in., the operator has approximately the same power over the lever at any point of its stroke. With the Turnover lever, however, there is a distinct loss of power on nearing the end of the movement whether moving towards normal or reverse, but the position of the lever near mid-stroke lends itself better to the exertion of power than any position of an ordinary lever. As previously explained, double wire compensation depends on the release or "escapement" of the cam wheels from the drive to the points or signals at each end of the stroke, so that the force required to complete the travel of the lever after the cam wheel has escaped is very much reduced and the (comparative) awkwardness of the lever's position is therefore not material; on the other hand the very long lever travel and the advantageous position of the lever about mid-stroke enable great power to be applied to the wires while the cam wheels are in engagement.

With ordinary single wire operation of signals the wires have to be comparatively slack when the signal is put to danger so that the balance weight at the signal post may be able to pull the wire back. The first portion of the lever travel in pulling off the signal is thus uselessly employed in taking up this slack. With double wire working, however, the wires are always in considerable tension, due to the compensator weights, so that no travel is lost in taking up slack.

With rod operation of points there should be no appreciable loss of stroke due to slack while the cranks and joints are new, but after wear has taken place this loss cannot be avoided, as all the joints and cranks which are in pull while the points are being moved in one direction are in push when they are moved in the other; so that if a run of rods has, say, sixteen pins between the lever and the function, a wear of $1/16$ in. on each pin will cause a loss of at least 1 in. of stroke, equivalent to about $1/5$ of the lever travel. This loss is emphasised if, as usually happens, an attempt is made to correct it by increasing the stroke at the adjusting crank near the points, for in that case the force to be transmitted by the rod is increased proportionately, and hence any tendency for the rods or cranks to buckle or spring is increased.

With double wire working, on the other hand, as wires are always in tension, any wear in wheels or pulleys will merely tend toward a slight increase in the frictional resistance, and no increased loss of travel should result

Owing to the greater tension in the wires, resulting in a comparative absence of slack, it is found that the wire pulleys for double wires can, with advantage, be spaced about twice the interval desirable for single wires, 20 yards being a usual figure. This results in a reduction of friction loss in the transmission as compared with ordinary signal wires and a still greater reduction as compared with rodding carried on rollers at 2 yards to 3 yards interval. As a consequence of the combined benefit obtained from decrease in friction, and the very long lever travel with no loss of stroke through slack, it has been found possible to operate ordinary double ended point connections by Double Wires up to a distance of about 600 yards from the lever.

Broken Wire Protection

The method employed to compensate for temperature changes in Double Wires results, as has been explained, in the wires being normally under a considerable tension, but as the tension in each of the two wires is equal, and tends to turn the cam wheel in opposite directions, their effects balance out and the cam wheel does not move. The breakage of one of the wires, however, would remove one side of balance and leave the tension of the other wire free to operate the cam wheel. This considerable energy stored up in the transmission which might possibly be released by breakage of one of the wires is one of the most important points which have to be considered by designers of this type of apparatus.

The difficulty has been overcome in various ways different designers, and Figs. 256 and 257 illustrate typical arrangements.

In the case of a Signal the breakage of the "pull off" wire would involve no risk, except possibly to the signalman, as the tension of the "pull on" wire, together with the energy stored up in the compensator weights (which are free to fall as soon as one of the wires breaks) would immediately restore the signal to danger if not already in that position. If, however, the "pull on" wire breaks, the tension of "pull off" would immediately rotate the cam wheel and clear the signal, but, in the absence of any "pull on" wire to check the rotation, the wheel over-travels, bringing the roller on the T-lever to the end of the groove where it inclines in toward the centre of the wheel (see Fig. 256), thus restoring the signal to danger.

With Points, no advantage would be gained by causing them to return to their normal position on the breakage of a wire; the object aimed at in this case being to insure that they do not move in either direction.

This is achieved in the design illustrated in Fig. 257 by means of a small double pawl located in a recess on the rim of the cam wheel, with which it normally rotates, being held in its recess by the equal tensions of the operating wires which are attached to either end of the pawl. When one of the operating wires breaks, thus freeing one end of the pawl, the pull of the other wire causes the pawl to spring out of its recess and engage with a fixed catch, thus preventing the rotation of the wheel. The double pawl and one of the fixed catches can be seen at the top left-hand corner of the cam wheel box in Fig. 257.

In addition to the safeguard just described, the mechanical detection of the points and bolt would prevent any possibility of their moving while the signal is at clear.

A somewhat similar double pawl is incorporated in the design of the Turnover lever illustrated in Fig. 260. This pawl is normally held clear of the square teeth on the segment by the equal tensions of the two operating wires, a reduction in the tension of one due to breakage allows the other to pull the pawl into engagement with the teeth, so locking the lever. This feature provides a means of drawing the signalman's attention to a broken wire, and also protects him from the risk of being injured by a lever with a broken wire, which might otherwise move violently over if the catch handle was lifted.

INDEX

RENASCENT BOOKS
*dedicated to creating facsimiles
and derivative works of
antiquarian & historical
value*

54698461R00126

Made in the USA
Charleston, SC
09 April 2016